W9-BJN-089

THE ECONOMIC
THEORY OF

Pollution Control

THE ECONOMIC THEORY OF
Pollution Control

Paul Burrows

The MIT Press
Cambridge, Massachusetts

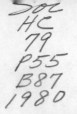

Soc
HC
79
P55
B87
1980

First MIT Press edition, 1980

First published in 1980 in Great Britain by Martin Robertson & Co., Ltd.

Library of Congress catalog card number   79–56534

ISBN 0 262 02150 1 (hard)
     0 262 52056 7 (paper)

Printed and bound in Great Britain.

Contents

ACKNOWLEDGEMENTS

I am indebted to David Pearce for allowing me access to some unpublished papers, and to Irene Waudby for her accurate typing and for easing the journey from manuscript to publisher.

for
Ann, Alison and Eileen Mary Frances Burrows

CHAPTER 1

Pollution and Economics

1.1 POLLUTION AS AN ECONOMIC PROBLEM

Environmental problems are not a new phenomenon. This is clear from the long history of attempts to control polluters, which in Britain goes back at least seven hundred years. Nevertheless, in the last twenty years there has come an increasing awareness that the magnitude of the pollution problem is sufficiently serious to warrant devoting greater efforts to studying the impact of pollutants and the means by which emissions might be curtailed. This is not to say that we can assert with confidence that in developed countries 'pollution is increasing', let alone that 'pollution is increasing exponentially', as some observers are inclined to do.[1] Undoubtedly the emissions of some pollutants are increasing, rapidly in some cases, but the picture is obscured by the fact that others remain constant or are even declining as far as can be judged from the data available. The emissions and concentrations of certain air pollutants in Britain, the United States and European countries may be constant or declining, in particular pollutants relating to coal burning in cities (smoke, sulphur oxides, particulates), while others are increasing, such as the hydrocarbons and nitrogen oxides emitted largely by vehicles. The changes in the quality of inland waterways also presents an uneven picture. In Britain the water quality of some rivers has improved, but nitrate concentrations resulting from farm fertilizer

1

run-off are increasing. Data on inland waterways are less comprehensive for other countries, but a similar picture of an uneven incidence between regions and a long term upward trend of pollution appears to apply. The rapid increase in the quantity of solid waste in the United States is probably typical of developed countries (though perhaps the U.S. experience is more extreme where it relates to non-toxic packaging materials). The quantities of toxic waste being dumped is hard to measure when significant amounts of it may be done secretly; but most observers express considerable concern and this is reflected in recent moves to legislate restrictions on the use of highly toxic chemicals.[2]

Whatever the trends in pollution levels, we are justified in treating pollution as a problem because it is clear, even from casual observation, that the current level of emission is higher (perhaps much higher) than it would be if the polluters had to bear all of the costs of their activities. It is this that lies at the heart of the economic analysis of pollution. Pollution is an economic problem partly because it reduces the value of some of the resources that society has at its disposal. 'Resources' here are to be interpreted broadly to include not only material resources such as labour, capital goods and raw materials, but also the facilities that the environment offers to people for recreation and as a healthy and pleasant place to live. The fact that pollution reduces the value of our resources is not, however, sufficient for it to constitute an economic problem. After all, for many types of pollutants we have available the technology largely to eliminate the major emissions. Pollution is an economic problem because it requires us to make choices, to resolve conflicts of interest; it is an economic problem because the means by which pollution can be reduced are themselves resource-using. For this reason much of the analysis in later chapters is concerned with a comparison of the costs associated with the *effects* of pollutants and the costs associated with the various methods of *abating* (preventing) pollution.

This is not to say that pollution is just an economic problem; that would be absurd. It is apparent that pollution is a problem for society that transcends the artificial boundaries of academic disciplines. It poses problems for all of the natural sciences, largely concerning the identification and measurement of physical impact of pollutants on the environment and the creatures, including man, that live in it. Pollution control challenges the engineer who is concerned with the design of abatement technology, such as sewage farms, and the lawyer who sees the law as a social instrument by which some form of restriction can be imposed on those who are polluting (see chapter 5).

In the final analysis society will resolve the pollution problem in a rational way only if it bases its response on the knowledge that each of the disciplines can offer. In this book we shall be concerned with the way in which economic theory can assist the resolution. In essence the theory has two things to offer. First, it can indicate the *type* of information that it is necessary to have if we are to be able to construct a balance sheet of the benefits of reducing pollution and the benefits of allowing pollution to continue. Second, it can attempt to predict the consequences of adopting alternative forms of pollution control policy. Economic theory does not, of course, in itself tell us what society should do to tackle the pollution problem. Rather, it attempts to provide a framework within which, if sufficiently good information can be obtained, a decision can be reached that at least is based on a non-partial view of the consequences of pollution control.

1.2   MATERIALS – FLOWS AND ASSIMILATIVE CAPACITY

Economists are accustomed to think in terms of the *flow* of goods and services in the economy, sometimes referring to the circular flow of income representing the interdependence of the demand and supply of goods and services.[3] They are less accustomed to accepting as an integral part of their analysis the interdependence between this flow of income and production and the natural environment. Yet the economic system both withdraws resources from the natural environment, and returns to the environment the waste materials generated by production and consumption activities.

The natural resources utilized by the economy are eventually returned to the environment in a variety of ways. Some are thrown out by the production process itself, such as waste gases and heat dispersed to the atmosphere and rivers; some result from the scrapping of old machinery; and others relate to the end of the economic system so to speak, namely the jettisoning of consumption goods when they have served their purpose. Only materials that can be recycled can be retained within the production–consumption flow; but recycling can never be 100 per cent so there is always a continuing leakage of materials back to the environment.

We shall, in chapter 2, talk about the damage to the environment resulting from the returning of these residual materials, but it is important to realize that not all emissions of residuals cause pollution damage. This is because the environment has a certain *assimilative capacity* to degrade the materials that it receives, and thereby to convert them into harmless, or even useful, forms. The best example of this is human waste, sewage, which at least up to certain concentrations is rapidly degraded by bacteria and made harmless, and is utilized as food by other organisms. Even crude oil in the sea is eventually incorporated into the ecological chain as food for bacteria. However, the assimilative capacity of the environment within a given time period is limited. What is more, some materials, such as the heavy metals, are not degraded by natural forces, and they accumulate in the environment when they are ejected by producers. Once the assimilative capacity is exceeded, detrimental effects may result from further emissions. In the analysis to follow in chapter 2, it will be assumed that pollution does exceed the assimilative capacity of the environment.[4]

1.3 AIM AND STRUCTURE OF THE ANALYSIS IN THIS BOOK

Books on the economics of pollution tend to fall into one of two categories. Either they minimize the economic theory content and concentrate on basic ideas and policy problems;[5] or they present the more sophisticated, up-to-date theory in the form of mathematical models.[6] The aim of this book is to cover most of the theoretical developments without resorting to mathematics. This makes the theory accessible to a wider range of economics students, and to researchers in other disciplines who are not *au fait* with mathematical economics.

Briefly the analysis falls into five parts. In chapter 2 the basic concepts that lie at the heart of the economic theory of pollution control are explained at some length. In the first part of chapter 3 the consequences of pollution for the efficiency of resource allocation *and* for justice are investigated. In the second part of chapter 3 we consider the possibility that the market system might achieve a solution to pollution problems in the absence of government intervention. In chapter 4 alternative government policy instruments for pollution control are analysed. Chapter 5 examines the role of the law in the implementation of pollution control.

CHAPTER 2

The Theory of External Cost

2.1 INTRODUCTION

In chapter 1 environmental pollution was described as an economic problem, that is to say a problem involving economic choices. The economist's explanation for the existence of pollution is the desire of firms and people to keep to a minimum their use of resources that they have to pay for. The theory that is based on this premise is known as the theory of external cost. In this chapter we examine this theory to provide a framework for analysing, in chapter 3, the consequences of pollution for the allocation of resources together with the obstacles to market solutions to pollution problems and, in chapter 4, the merits and limitations of a variety of pollution control policies. The immediate aim is to present the simplest analytical framework capable of handling a problem that, in reality, is highly complex. No attempt will be made to survey the extensive theoretical literature, which sometimes has allowed a taste for fine definitional distinctions and for analytical elegance to obstruct progress towards tackling the difficult evaluation of pollution and control policies in the real world.[1] Nevertheless the analysis must begin from some careful definitions of the terms that will be used frequently.

6

2.2 Some Definitions

While definitions are a matter of convenience, confusion will be minimized if we keep as close as possible to conventional usage. To begin with, the effects of pollution may well be viewed as costs to society of the kind generally referred to as *external costs*. Let us state a definition of external cost that is appropriate to an analysis of pollution and then elucidate its meaning:

> An external cost exists when a production or consumption activity induces a direct loss of utility, or an increase in production cost, which does not enter the decision calculus of the controller of the activity.

Three elements in this definition require explanation. In the first place the term 'cost' is to be interpreted broadly to include all adverse effects, but herein lies an important judgement as to which effects are 'adverse'. Is *any* pollution-induced change in any component of the environment adverse? If not, who is to draw the line between adverse and benign (or even beneficial) changes? Some ecologists argue that all physical changes should be regarded as costs, but an alternative answer is to assert that a change is adverse if human beings think it so. The latter view implies, for example, that the destruction of a species of insects that people do not care about (or even are glad to see disappear) is not regarded as being a part of the damage costs. This egocentric human view may well be the only operational definition, given that it is human beings alone who control the source of the problem. It is worth noting moreover that the increasingly fashionable analytical treatment of pollution in terms of materials–flows and assimilative capacity[2] in no way resolves the problem of distinguishing adverse from benign effects. The advantage of the materials–flow approach is its

emphasis on the interdependence, in terms of physical
flows, of the environment and economic activities. But
the description of an effect of pollution as 'adverse' implies
a *valuation* (not necessarily in terms of money) over and
above the identification of physical magnitudes. The establishment of priorities in pollution control, in the context
of a multiplicity of resource-using government activities,
necessarily involves such a valuation whether explicit or
implicit. But this is not to deny that the information
derived from a materials–flows analysis (such as the prediction of the physical consequences of an excess of pollution load over assimilative capacity) would be a prerequis-
ite of a complete valuation of pollution impact.

The second point about the above definition of external
cost is that it concentrates our attention on costs that
not only fall on people or firms outside the polluting
activity but also lie outside the decision calculus of the
polluter. Such costs will occur if the polluter is prepared
and allowed, by law, to ignore them.[3] In contrast to
normal market transactions, the polluter is able to ignore
the flow of external costs because there is no corresponding
(equivalent) flow of payments by the polluter to induce
him to enter the costs into his decision calculus (which
is the weighing of the costs and benefits to him of different
levels and forms of his activity).[4] It is this asymmetry
of cost and payment flows that is the crucial element
as far as the analysis of pollution is concerned. It dis-
tinguishes the cases we shall refer to as external costs
from other forms of interdependence which have some
characteristics in common. For example, the use of labour
by a firm imposes disutility on workers analogous to
the cost imposed by pollution on those affected, but this
is not an external cost because the obligation to pay
a wage at least equal to the disutility internalizes the
cost to the employer. In a slave society, on the other
hand, workers do not own their labour services and the

involuntary employment of slaves does impose an external cost on them. Notice that we do not identify external costs simply with those costs that are ignored in voluntary, market transactions. This is because the flow of equivalent payments could be in the form of compulsory taxes. We can then speak of external costs being internalized either by voluntary transactions between polluters and pollutees or by tax payments; and we can describe any damage costs that remain after the polluter has responded by adjusting his activity as *internal* costs, in the same sense that the loss of utility to workers is an internal cost in a normal labour market. Thus costs are external to the polluter's decisions if there are no internalizing payments of any kind, market or non-market.

The third element in the definition of external cost hinges on the phrase 'a *direct* loss of utility or increase in production cost'. There are many forms of interdependence between activities in a market system that we do not wish to include in the term 'external cost' for the purposes of environmental economics. In particular it is simplest if those interdependencies that are transmitted indirectly through changes in the prices of goods or factors of production are omitted.[5] If, for example, there is an increase in the demand for bricks, owing to an increase in factory building, which raises the price of new houses, then buyers of new houses will suffer a loss of real wealth. But effects of this kind are not external to the decision calculus of those undertaking new factory building. In bidding up the price of bricks they are obliged to take account of the demand for bricks for house building, and by implication take account of the utility loss to those now unable to obtain houses at the previous cost. Only if the value of bricks in new factory building exceeds (at the margin) that in house building will the price of bricks be bid up. In general, changes in the allocation of resources between uses in conjunction with changes in relative prices will have distributive consequences for different groups in society; but the distinguishing characteristics of the *direct* interdependencies of the kind we are calling external costs is that the market system may not reflect these effects. An increase in the demand for river space for the effluents from new factories may not involve the factory builders in bidding against other demanders of river space for effluent disposal and against demanders of clean rivers. With the river and its assimilative capacity available free of charge there is nothing in their decision calculus to induce the factory-builders to compare the value of the river *to them* with the value of the river *to others* (including the value of a clean river to the public in general). The impact of effluent disposal is direct, not via price changes, and is truly external to the decision-taker. The result of the external cost may well, as we shall see, be a reduction of the productive potential of the economy's resources, but there is nothing *technically*[6] different between such external costs and the types of interdependency we have excluded. After all, any external cost can become an internal cost if the incentive structure within which polluters operate is altered.

Some economists have sought further to limit the class of interdependences that are to be treated as external costs by requiring that the *deliberate* imposition of costs on others be excluded.[7] The motive for this limitation appears to be the desire to reserve for separate treatment cases where (say) individual A imposes a loss of utility on his neighbour B by playing his transistor radio in the garden and, what is more, enjoys doing so. One can see why cases of this kind, as well as the criminal offences of murder, rape and robbery, require separate consideration, because here the intentions of the perpetrator of the offence are central to the analysis. However, a deliberate/non-deliberate distinction raises problems in the context of environmental external costs. It is apparent that in practice, many pollutants derive from the normal, everyday production activities of polluters. Clearly these acts of polluting are deliberate in the sense that they are the predictable (and in some cases certain) outcomes of production with existing processes, and it would be nonsensical to exclude their consequences from our definition of external costs. It is a nice, almost metaphysical, question whether the fact that a polluting activity is deliberate implies that the consequent imposition of costs on others is also deliberate. Is the owner of a factory that is emitting smoke not only deliberately polluting the atmosphere but also deliberately imposing costs on people five hundred miles away? Rather than pursue the possibility of deliberate pollution leading to unintentional harm, we shall simplify the situation by assuming that in general polluters do not pollute in order to impose costs on others, but rather to avoid the costs of not polluting. This seems reasonable because we shall be concerned mainly with pollution by firms, and as far as the decision to pollute is concerned, malevolent intentions are perhaps less plausible for firms than they are for people. The imposition of costs will therefore be viewed as an incidental consequence of the cost-minimizing behaviour of polluters; but nevertheless a distinction will be maintained between deliberate and accidental polluting *events*. It will prove to be important for policy purposes to distinguish, for example, the persistent emission of chemicals into the atmosphere owing to normal production in a chemical factory from the emission resulting from an explosion owing, for example, to the unnoticed corrosion of chemical carrying pipes. Similarly, the problem of limiting the dumping of oil residues during normal tanker operations at sea is likely to differ from that of controlling the risk-taking activities of tanker captains which lead to collisions at sea and consequent oil spillage.

External costs can be viewed as a wedge between the private, or internal, costs of a polluting activity which are incurred by the polluter and the total cost of the activity to society, which we call the *social* cost. We turn now from matters of definition to the causes and effects of external costs.

2.3   ABATEMENT COSTS AND POLLUTION COSTS

2.3.1   *Cost-minimizing polluters and methods of abatement*

At first sight the act of polluting may seem to be the consequence of irrational behaviour by the polluter, such as an inability to foresee the eminently predictable consequences of his activities. It may well be that some pollution is the result of myopia, but polluting is so diverse and widespread that one must seek an explanation of such behaviour that accepts that it is rational in terms of the motivation and obligations of the polluter. The explanation that will be discussed here requires the assumption that polluters, whether producers or consumers, attempt to minimize the costs that they incur in their production or consumption activity. As far as firms are concerned this cost-minimization assumption is weaker than the conventional assumption that firms are profit-maximizers, because firms pursuing other objectives than the earning of profit (such as the maximization of sales revenue) are also likely to attempt to minimize their cost per unit of output.

In order to minimize the cost per unit of its activity a firm will compare the costs that it would incur if it polluted freely and those that it would incur if it controlled (abated) its pollution, and will choose the lower of the two. The costs that the firm faces as a result of pollution will depend on the degree of freedom to ignore costs that the law allows, and on the penalties that it imposes on polluting.[8] The costs to the firm of controlling pollution, on the other hand, can be thought of as the *benefit* to the firm from polluting. These costs, which we shall call polluter-abatement costs, are the costs of modifying the size *or* character of the polluter's activity, which either reduces the quantity of pollutant emitted or alters its composition (for example reduces its concentration or slows its speed of emission) in such a way that the impact on those affected is reduced.

One of the curious features of the economic analysis of pollution for many years has been the pervasive assumption that if the amount of firms' pollution is to be reduced then their *output* must be cut. In reality the quantity and quality of emissions depend not only on the level of output, but also on the nature of the production process employed. Consequently we can distinguish between the costs associated with two alternative methods of abatement.

(1) *The cost of cutting output*

If a firm feels committed to a particular production process, as it may well do in the short run at least, a reduction in the amount of effluent will involve a sacrifice of output.[9] The cost to the firm in this case is the loss of profit on the units of output eliminated. In a neoclassical full-employment world the cost to society of undertaking abatement of this kind also is the loss of profit to the firm: all factors of production laid off can be re-employed as productively elsewhere. Once the risk of the non-redeployment of factors that are laid off is admitted, however, the social cost of abatement must include the value of the output attributable to those laid-off factors that cannot find an alternative use. For the purposes of the analysis that follows, however, it will be assumed that factors do have alternative uses, which is generally true in the long run at least, so that the abatement cost is derived from the firm's profit curve. This is illustrated for a simplified model of the firm in figure 2.1, where in quadrant (a) the downward-sloping curve is the profit obtainable

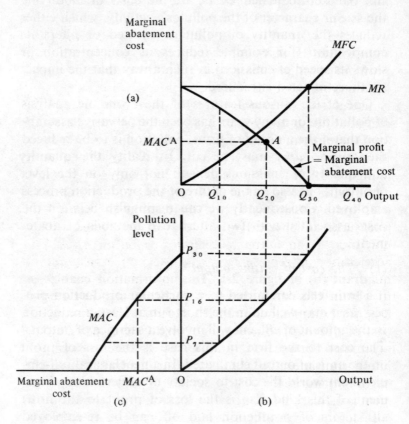

FIGURE 2.1 *Marginal cost of abatement through cuts in output*

from successive extra units of output, being the difference between marginal production cost, *MFC*, and marginal revenue, *MR*, at each output. Once output Q_{30} is reached, total profits (the area under the marginal profit curve) are maximized; marginal profit is zero. If pollution is abated by cutting output from, say Q_{40} to Q_{30} no loss of profit occurs and marginal abatement cost is zero. Further abatement by reducing output to Q_{20} incurs the loss of profit represented by the triangle Q_{20} A Q_{30}, and if one extra unit of output were eliminated the cost would be MAC^A. It is clear that for movements from right to left the marginal profit curve shows the marginal profit *loss* owing to abatement; that is, the curve represents the marginal cost of abatement through output cuts. It is useful to extend the analysis a little to obtain the marginal abatement cost for different levels of pollution (and therefore of pollution abatement) rather than output. If pollution is linearly related to output, then the curve in figure 2.1 (a) simply converts to the required MAC curve with pollution on the horizontal axis. However it is quite possible that the increase in pollution level (measured in physical terms) that is associated with a unit increase in output accelerates as output rises. The acceleration is represented by the exponential curve in quadrant (b) of figure 2.1. This information enables us to obtain the MAC curve in quadrant (c). For output Q_{20}, for example, pollution P_{16} results and the (marginal) cost of abating the P_{16} unit of pollution is MAC^A. The postulated nonlinear relationship between output and pollution translates the MAC curve in quadrant (a) to the curve in quadrant (c). This latter curve says that as abatement is increased, i.e. as output is cut further, the marginal cost (loss of profit) increases for a given size of pollution cut. One reason for this is that initially a 1 unit output cut leads to a large reduction in pollution, but as production declines a unit cut becomes less successful at abating pollution because the lower output levels are not the heavy polluters. Thus, a 10 unit reduction in output from Q_{30} to Q_{20} saves 14 units of pollution, but the fall from Q_{20} to Q_{10} saves only 9. This is one reason why savings in pollution of a given size require increasingly large sacrifices of output and profit as the firm moves towards total abatement, i.e. zero pollution. The other reason is that the profit loss for an extra unit of output sacrificed increases as output falls, because with rising marginal cost the initial units of output yield more profit than the extra ones (so the curve in quadrant (a) is downward-sloping).

The cost to the firm of abating through output cuts is dependent on the demand for its product. An increase in demand raises marginal revenue and (for a given plant size and marginal cost curve) shifts up the marginal abatement cost curve in figure 2.1(a).[10] The benefit to individual polluters obtained from polluting therefore depends on demand, and the argument has been extended to suggest that economic growth is characterized by rising demand and consequently by an increasing incentive to pollute. But does the demand argument, which undoubtedly can be applied to individual polluters, also hold for polluters in the aggregate? Does growth necessarily lead to an increasing *aggregate* demand for the use of environment as a dustbin? As long as any technological innovation involved in the growth process is not of the kind that alters the pollution produced by a given level of output, and therefore leaves the pollution/output curve in quadrant (b) unaltered, the upward trend of profitable output must shift up the aggregate *MAC* curve. This is true whether the basic cause of growth is higher investment, or labour- or capital-saving innovation, or both. It is becoming clear at this point that the choice of production processes (technology) is central to the implications of growth, or even of a particular quantity of output, for the environment, and we shall examine the relationship between production process and abatement cost in a moment.

While the analysis of abatement through a reduction

in output clearly applies only to polluting firms, an analogous line of reasoning can be used for polluting consumers. The characteristics of a consumption activity that determine its external effect are the nature of the good used, for example the emission-saving efficiency of a car engine at a particular speed, and the manner of its use, such as the speed at which the car is driven. If these characteristics are fixed then abatement through a cut in the level of consumption induces a loss of utility similar to the pollution-abating firm's loss of profit. As long as the marginal utility from consumption declines as consumption increases, the marginal abatement cost curve will be similar in shape to that in figure 2.1 quadrant (a) or (c). It will simplify the subsequent analysis considerably if we recognize that the consumers' utility losses and firms' additional costs that result from pollution abatement are similar, in as much as they both comprise a component of society's evaluation of the *benefit* of a polluting activity. This saves us having to repeat the analysis for polluting firms *and* polluting consumers, abatement costs being assumed to be incurred by *all* polluters who abate.[11]

(2) *The cost of changing production processes*

When you look closely at a variety of polluting production activities in the real world one feature stands out: the pollution often derives from the use of a production process that is more polluting per unit of output than alternative *known* processes. This means that, technically, abatement by the polluter need not rely on a reduction in the level of the polluting activity or on the future invention of an abatement process.

Output is produced by utilizing inputs in a particular combination, that is by employing a particular production process. A process that employs men utilizes the labour input; similarly, a process that creates pollution utilizes the environment which is in the clean form acceptable for other purposes. So we can say that such a process utilizes the *environmental input.*

A process is more or less environment (or pollution)-intensive according to the size of the ratio of the environmental input to the inputs of labour and capital.[12] In figure 2.2 the straight lines (rays) from the origin, $P1$,

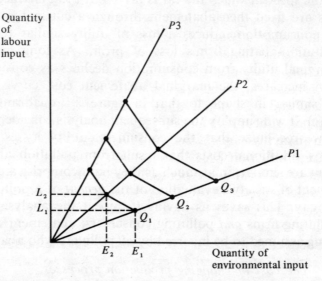

FIGURE 2.2   *Production processes and the environment*

$P2$, $P3$, represent production processes that offer alternative means of producing outputs Q_1, Q_2, Q_3. For example output Q_1 can be produced by the input of E_1 of the environment and L_1 of labour if process $P1$ is employed, or by E_2, of the environment and L_2 of labour if process $P2$ is used.[13] A switch from process $P1$ to $P2$ represents the adoption of a more labour-intensive, less environment-intensive (i.e. less-polluting) process. A switch of this kind may involve, for any output, either a reduction of the total *quantity* of pollutants emitted or a change in the *form* in which they occur. This reflects the fact that reference to the amount of pollution (e.g. 'less-polluting') is really a shorthand for an index of quantity and quality (e.g. the form in which the emission occurs) of pollution.

A process switch that reduces the quantity of pollutants emitted may involve either the introduction of recycling or the adoption of a technology requiring smaller quantities of inputs of materials. Recycling feeds back part of the output of waste materials into the production process, whereas materials-saving reduces the throughput of materials. Examples of recycling abound: many metals are recycled in some degree at present – iron, copper, tin and lead for example – and so are varying proportions of paper, rubber and glass.[14] Materials-saving may come from an increase in efficiency in the utilization of materials in the production process, such as the reduction in the quantity of coal used per unit of electricity produced in modern power stations, or from changes in product design, for example the use of thinner body panels in cars.

Processes that change the *form* in which the pollutants are emitted are typically devices to collect pollutants at the point of exit from the production process. Thus electrolytic extractors or filters in chimney stacks may collect soot, grit or fibrous materials, and convert the materials otherwise dispersed into the air into more compact solids, which must then be disposed of on land or in water. Of course 'the amount of pollution' resulting is reduced in total only if a means of disposal can be found that is less damaging that the air pollution.

Process switches are not normally a costless means of abating pollution (reducing the environmental input, E). If E is to be reduced, usually the input of other factors, labour in the simple analysis, must be increased. The extra units of the other factors are required to construct, install and operate the new production process.

It is apparent that we should be able to identify a curve showing the marginal cost of abating through changes in process, which can be compared with the marginal cost of abating through cuts in output in the search for the least cost method of polluter abatement. Let us approach the problem rather obliquely by first viewing the marginal cost of abating through output cuts from the point of view of the polluting firm's input choice. Imagine the competitive firm represented in figure 2.3 (a); it is initially utilizing the environmental input up to a level, say E_{10}, which is associated with the profit-maximizing output level Q_{110} if process $P1$ is employed at A. If Q_{110} maximizes profit, then point A is the most realistic starting point because $P1$ is the most polluting process, the one that extracts the most savings of other

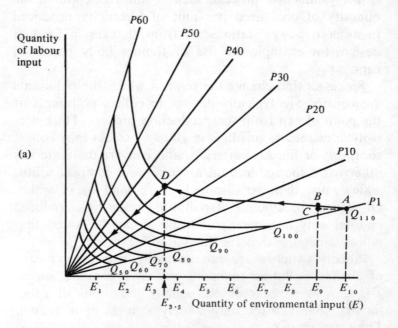

FIGURE 2.3   *Output cuts, process switches and abatement costs*

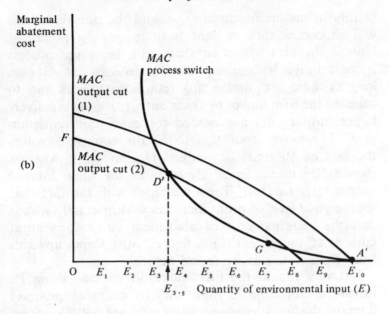

(b)

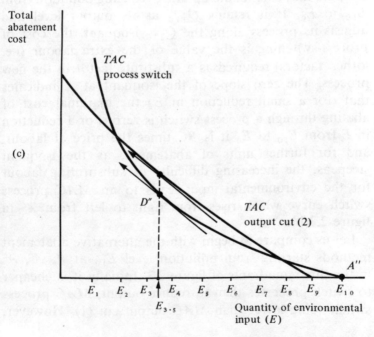

(c)

factors to maintain output Q_{110}, and the one that firms will choose if they do not have to pay to pollute.[15] Now if the firm wishes to abate with the same process it must move back along the $P1$ process ray. But as long as there are diminishing returns to factors and to scale, as the firm moves to lower output levels successively larger output cuts are needed for each unit reduction in E. The move from $E_{5.5}$ to E_4 for example involves the loss of 10 units of output (Q_{90} to Q_{80}), whereas the smaller move E_4 to E_3 requires the same loss of output (Q_{80} to Q_{70}). This, combined with the fact that the *marginal* level of profit increases as output falls, means that the marginal cost of abatement curve for output cuts, MAC output cut (1) in figure 2.3(b), slopes upwards from right to left as we found before.

But why should the firm move downwards along $P1$ instead of adjusting its input mix by changing process? Suppose the firm is contemplating reducing pollution from E_{10} to E_9. If it retains Q_{110} as its output level and adjusts its process along the Q_{110} isoquant, the cost of process-switching is the value of the extra labour (i.e. 'other' factors) required as a substitute factor in the new process. The zero slope of the isoquant at A indicates that (for a small reduction in E) the marginal cost of abating through a process switch is zero. For a reduction in E from E_{10} to E_9 it is BC times the price of labour, and for further units of abatement, as the isoquant steepens, the increasing difficulty of substituting labour for the environmental input leads to an MAC process switch curve which rises from right to left from A' in figure 2.3(b).

Let us compare to begin with the alternative abatement methods starting from pollution level E_{10} at A, A', A'' in the three quadrants of figure 2.3. Initially it is cheaper to switch processes than to reduce output, MAC process switch being lower than MAC output cut (1). However, as the process-switching takes place the firm's marginal cost curve rises, owing to the extra resources used for the switch, and the level of profit on each unit of output falls. Consequently the MAC output cut curve falls, say to the position of MAC output cut (2) in figure 2.3 (b). Once the level of pollution $E_{3.5}$ has been reached through process-switching, at D, D', D'', there is a cross-over of the MAC curves (the *total* cost curves in quadrant (c) being parallel at this point) and it becomes cheaper to leave the technology at process $P30$ and thereafter abate by reducing output.[16] The isoquants in figure 2.3 (a) yield this cross-over of costs: initially they have a small slope, starting from A, but the slope increases rapidly as process-switching occurs at a given output. As isoquant Q_{110} in figure 2.3(a) approaches the vertical, marginal process-switching costs approach infinity as the MAC process switch curve in quadrant (b) shows.

The path of adjustment that results is a change of technology from point A along the isoquant Q_{110} until process $P30$ is reached at point D, which corresponds to the cross-over point D' in figure 2.3(b). Thereafter further abatement comes from cutting output along the $P30$ ray, as shown by the arrowed adjustment route in figure 2.3(a). The cost-minimizing firm will treat the lower segments of the MAC output cut and MAC process switch as its overall MAC curve, $FD'GA'$ in quadrant (b). Notice that the marginal cost of abating through a cut in output is higher starting from point D than it would have been from A. The reason is that the move from E_{10} to E_9 along $P1$ involves a smaller output cut than does the move from $E_{3.5}$ to $E_{2.5}$ along $P30$. This is the result of the lower marginal productivity of the environmental input when $P1$ is used.

A different configuration of the isoquant map could produce a cross-over in the opposite direction. If the isoquants were of medium slope throughout then the

MAC process switch curve could initially (moving to the left from A' in figure 2.3(b)) lie above the *MAC* output cut curve, but the constant slope would allow the more rapidly increasing marginal cost of cutting output eventually to rise above *MAC* process switch. This situation (not shown in the figures) would induce the firm to abate from A initially by cutting output and eventually by altering its technology.

The position of the least-cost-abatement curve is not fixed for all time but depends on technological knowledge and economic conditions. If, for example, polluting industries experience booming profits during a period of rapid growth, the cost of abating through output cuts increases; the *MAC* output cut curve for a firm, in figure 2.4, shifts from position (1) to position (2). The least-cost-of-

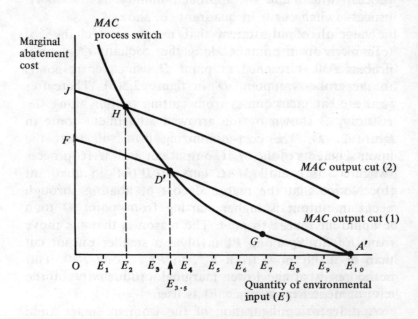

FIGURE 2.4   *A change in the cost of abating through output cuts*

abatement curve shifts from $FD'GA'$ to $JHD'GA'$, which means that for E levels below $E_{3.5}$ the lowest cost of abatement has increased. If the firm had previously decided to abate, partly by cutting output, from E_{10} to a level of E below $E_{3.5}$, then it will now reduce the *level* of abatement in line with its increased cost. If, on the other hand, the firm had previously abated only to a level above $E_{3.5}$ by switching processes, because the lowest marginal cost of abating units of E above $E_{3.5}$ has not altered (the cost of process switches remaining constant), the chosen level of abatement will not change. However, the abatement of the units of E in the range E_2 to $E_{3.5}$, which were previously abated by cutting output, is now achieved through a process switch. Raising the value of output leads the firm that is abating down from E_{10} to continue switching process until E_2 is reached instead of changing to output cuts at $E_{3.5}$ as it did before the value of output increased.

To work out the implications of a decline in the cost of abatement through cutting output it is necessary only to reverse the arguments in the previous paragraph. But what are the consequences of innovations in abatement technology? Innovations of this kind make it easier to substitute labour ('other factors') for the environmental input. This is reflected in a flattening of the isoquants in figure 2.3(a) and in a downward move of the MAC process switch curve in figure 2.5. The least-cost-of-abatement curve moves from $FD'GA'$ to $FKLA'$ and it should be apparent that, if the firm were abating from E_{10} to a level of E above $E_{3.5}$, the technical innovation would increase his level of abatement because the least-cost method of abating is now cheaper at the margin than before. Moreover, if abatement is taken to below $E_{3.5}$, process-switching becomes the lowest (marginal) cost method of abating in the E_1 to $E_{3.5}$ range. However, if abatement were taken below E_1 before the innova-

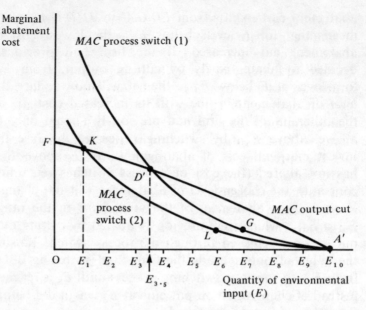

FIGURE 2.5   *A change in the cost of abating through process switches*

tion, the *level* of abatement would not alter (output cuts being used at the margin) but the extended use of process switches would still apply.

One dimension of the problem of finding the cheapest form of pollution abatement which has not yet been mentioned is the possibility of economies of scale being achieved through the centralization of the treatment of pollutants. We have been assuming that the firm, if it chooses to abate by altering the relationship between output and pollution levels, will alter its *own* production process. But in reality the choice is more complicated because there may exist several ways of reducing the level of pollution per unit of output. In a simple world in which the average cost of treating pollutants is independent of the quantity treated, that is to say where there are constant returns to scale in treatment, any polluter large or small can abate at the same cost per unit if treatment at the end of the production line is the cheapest method. To take an extreme example, under a regime of constant returns to scale in treatment each household could treat its own sewage as cheaply as a centralized treatment facility could do the job. In reality, of course, the capital-intensive nature of the treatment of pollutants means that the spreading of capital costs over large quantities of treatment will produce substantial economies of scale. As a result centralized, out-of-factory (or home) treatment is competitive at least for the more common types of industrial and domestic liquid and solid wastes. From the point of view of, say, an individual polluting firm the marginal cost of effluent treatment in a centralized facility is given and constant. Referring back to figure 2.3(b), if a horizontal MAC centralized treatment curve were inserted and it lay below point D', the least cost method of abating to $E_{3.5}$ would be partly a change of process and partly centralised treatment.

The discussion of process-switching as a means of breaking the link between the levels of output and pollution has naturally concentrated on the behaviour of polluting *firms*. However it is not difficult to provide an analogous line of reasoning for polluting *consumers* by using Lancaster's theory of consumer choice.[17] Lancaster's theory is based on the presumption that consumers buy goods not because they like goods as such, but because the goods have characteristics or attributes (such as the flavour, texture and nutritional value of beef) that yield utility. A particular good may have more than one attribute (as in the case of beef) and a particular attribute (e.g.

nutritional value) may be offered by more than one good. The consumer is concerned to maximize his utility, which is positively related (with different weights, no doubt) to the various attributes of goods that he enjoys, subject to his income constraint. This approach to consumer choice offers persuasive explanations of product differentiation, the development of new goods and a number of other readily observable features of markets in developed economies.[18] For our purposes an application of the theory to the analysis of abatement costs can begin by postulating that, in our simple world, goods have only two attributes that enter *somebody's* utility function,[19] and there are three goods available that possess these attributes in differing degrees.[20] The two attributes are comfort from domestic heating and lack of environmental impact (the inverse of pollution intensity); and the three goods available are ordinary domestic coal, smokeless fuel and gas. Measuring the two attributes on the axes, comfort vertically and lack of adverse impact horizontally from left to right (zero being the neutral case), the three goods are represented in figure 2.6 by the three rays from the origin. Coal is presumed to offer (per unit of cost) a high degree of comfort owing to the combination of the attractiveness of an open fire and moderate efficiency in heat generation, smokeless fuel is less successful in terms of comfort, and gas is least successful. The reader will perceive that in order to limit the example to two attributes we are interpreting the comfort attribute as a weighted index of heat provision and attractiveness to the eye, with the latter being more heavily weighted. Radiator-loving readers can substitute their own ranking of the three goods with respect to the comfort attribute. When the environmental neutrality attribute is considered we find the ranking is reversed: for a given level of comfort coal is the heaviest polluter of the air, smokeless fuel the next and gas the least

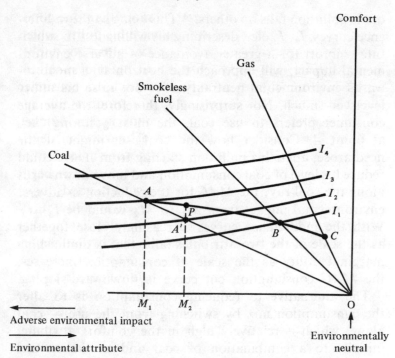

FIGURE 2.6   *Abatement of pollution from consumption activities*

polluting. The rankings in terms of the two attributes imply that if the consumer spends the whole of his heating budget on coal he can reach A (maximum comfort, maximum pollution); if all is spent on smokeless fuel B can be reached; if all on gas C can be reached. By sharing the budget between coal and smokeless fuel (say heating some rooms with each) or between smokeless fuel and gas, points along AB and BC respectively are attainable.

Now it is reasonable to assume that the average consumer will derive relatively little utility from *his own* fuel being environmentally neutral given that most of the cost of his pollution falls on others.[21] This being so the indifference curves, I_1, I_2 etc., describing his willingness to substitute comfort for a greater avoidance of adverse environmental impact, will approach the horizontal: a move towards environmental neutrality does not raise his utility level very much. Not surprisingly, therefore, the average consumer prefers to use coal, the most polluting fuel, at point *A*. Consider how the coal-user might, if the need arose, abate his pollution starting from *A*. He could reduce his level of coal consumption and move downwards along the coal ray; the *MAC* for the reduction in adverse environmental impact from M_1 to M_2 would be $I_3 - I_2$. With the indifference curves increasingly close together as the scale of the two attributes falls, that is diminishing marginal utility as the scale of consumption increases, the *MAC* consumption cut curve is downward-sloping.

The alternative to reducing consumption is to alter the consumption mix by switching from the use of coal alone, which is relatively high in the comfort attribute, initially to a combination of coal and smokeless fuel on *AB*, then to smokeless fuel alone at *B*, and on to smokeless fuel and gas combinations on *BC* and eventually to gas alone at *C*. But this progression involves the loss of utility owing to lower levels of the comfort attribute being achieved because *AB* and *BC* are downward-sloping; that is, there is a trade-off between the two attributes. Notice, however, that the loss of utility for a given degree of abatement, such as from M_1 to M_2, is less if the consumer switches between types of fuel, from *A* to *P*, than if he is constrained to using coal and reduces consumption, from *A* to *A'*. The reason is that by switching to another type of fuel the individual can partially protect his attainment of comfort while abating.

The choice between reducing consumption and switching to goods with different attributes is not so conceptually simple if goods have multiple attributes, as we can illustrate by a brief examination of the use of transistor radios in public places. Transistor radios offer continuous entertainment for those who find a state of nature too demanding. When used in public they also provide continuous annoyance for those who seek quiet. The audibility of the radio to the carrier is not much affected by low-cost earphones (usually included in the purchase price) which eliminate the nuisance to other people, except to those who object to seeing the flower of the country's youth twitching to soundless music. Why then does it seem so rare that the switch to radios-with-earphones is made? There are at least three possible explanations. First, perhaps transistor radio users *like* causing a nuisance so that the nuisance becomes another attribute (to the user) of radios without earphones. In this case the marginal abatement cost for earphones is the loss of the nuisance attribute. Second, perhaps transistor radio users like to provide a public service, failing to perceive that to some this is a disservice. Presumably users motivated in this way will desist when confronted with forthright assertions from sufferers that the service is not required. Third, perhaps the problem lies in the fact that the appreciation of canned music is a social activity enjoyed more by people in a group than by a person on his own. Earphones may impose technical, not to say physical, constraints on the enjoyment of this social activity so that their use incurs the loss of this social attribute. In this example, as in all cases of external costs derived from consumption activities, there is room for wide disagreement about the subjective evaluation of the attributes of the consumption involved. It is imaginable that, depending on the motivation of transistor radio users, the claimed abatement cost of using earphones may be regarded by many in society as illegitimate (case 1 above), misconceived (case 2) or positive and to be weighed against the value of the external cost (case 3). Depending on the attitudes of transistor radio users, abatement by altering the consumption technology (attribute mix) may or may not be preferable to them to cutting volume or altering the location or the amount of radio listening. In conclusion, it is apparent that the environmental impact of the consumption of a good depends on the amount of consumption and on the inherent characteristics of the good (product design and the way the good is used), and the cost of abating pollution from consumption activities depends on which of these variables is altered in the abatement procedure.

2.3.2 *Pollution costs*

Polluter abatement costs are the costs of reducing the amount (i.e. quantity and/or concentration) of pollution and they represent the *cause* of pollution by a cost-minimizing polluter. We turn now to the *effects* of pollution, the costs to the rest of society which we shall call pollution costs. Before considering (in sections 2.4 and 3.1) the importance of the legal system in determining the extent to which pollution costs are also *external* costs as far as the firm is concerned, let us explore a little the main components of pollution costs. 'Once pollution has occurred, owing to the failure of the polluter to abate, two types of cost may be incurred by society. In the first place, if there are no mitigating responses by other members of society, the cost incurred will be the reduction in the value of the consumption and production activities that are affected by the pollutants. These will be referred to as *pollution damage costs*. The variety of damage costs is considerable, ranging as they do from the tangible, measurable loss of profit by firms whose production costs are raised by air or water quality deterioration to the personal sensitivity of a Mozart lover whose enjoyment of an aria from *The Marriage of Figaro* is impaired by a passing aeroplane. In general, increases in production costs owing to pollutants are the more objectively measurable simply because the extra factors necessary to offset the impact are valued in the market. However, the detrimental consequences of some pollutants for the mental and physical well-being of people may be measurable in money terms, and clearly such health effects influence people's consumption activities as much as their efficiency as factors of production. Leaving aside the theoretical and practical difficulties of measurement, we can say that reductions in the value of production and consumption activities are *conceptually* similar.[22] Both of them reduce, either directly or indirectly, the utility derived by society from its available resources. Thus, the value of the broadcast of *The Marriage of Figaro* from the resources available to the *BBC* is directly reduced by extraneous noise; and a firm that has to use extra detergent in a cleaning process owing to the deterioration in water quality is being forced by the pollution to take up extra factors which have alternative uses elsewhere in the production of utility-yielding goods.

The second type of pollution cost arises from defensive operations to reduce damage costs, and will be called *damage reduction costs*. While polluter abatement costs are the costs of reducing the amount of pollutant emitted, damage reduction costs are the costs of reducing the damage resulting from any given amount of pollution. The latter can be of several kinds which are not always distinct in practice, such as alterations to the size, form and location of pollutees' activities to minimize the impact of pollution on them, attempts to arrest the spread of the pollutant once emitted, and cleaning-up operations. Some of these costs may be incurred by individual pollutees. A cost-minimizing pollutee will react to damage costs imposed on him either by simply bearing the cost, or by reducing his activity level (the pollutee equivalent of the polluter's output or consumption cut), or by altering his production or consumption process to make it less sensitive to the pollutant (the equivalent of the polluter's process-switching), or by moving away, whichever is the cheapest. Other damage reduction costs may be borne by the taxpayer, for example through the additional public expenditure on the purification of water for drinking owing to river pollution, and through public clean-up operations for litter, oil spills and toxic chemical wastes dumped by firms, where these costs are not directed back to the polluter.

For the most part it is polluters who can influence the amount of pollution, and the rest of society who can alter the amount of damage resulting from a given emission. But this is not a clear-cut division, because the polluter can be involved in damage-reducing activities such as clean-up, and the polluter could relocate his polluting production (or consumption) to reduce the damage consequences of his emissions. This polluter relocation option adds another dimension to the cost-minimization calculation. The polluter may reduce output or switch processes to reduce pollution; or damage may be reduced by the pollutees' or polluters' relocating, or by pollutees' altering their activity in its present location. This choice is complicated further by the availability of many combinations of these alternatives. In order to maintain his sanity the reader would do well to bear in mind the fact that the least-cost form of abatement by the polluter provides the abatement cost curve, and the choice of the smallest of damage costs and the various damage reduction costs determines the pollution cost curve.

The various types of pollution cost will frequently be treated together in the analysis to follow, and pollution costs will be assumed to display certain characteristics in the aggregate. First, in much, but not all, of the analysis the relationship between total pollution cost and the amount of pollution[23] will be assumed to be positive and continuous (that is it involves no discontinuities or 'steps'), as shown in figure 2.7(a). Much of the literature on external costs also assumes that the increase in total cost associated with a given increase in pollution rises as the amount of pollution (environmental input) rises; that is to say that marginal pollution cost is a positive function of the amount of pollution, as shown in figure 2.7(b).[24] This is not, however, a necessary assumption and, as a number of economists have been at pains to point out, it is not necessarily accurate as a description of serious pollutants: it is possible that after a point the total pollution cost curve becomes flatter as pollution increases, so that the marginal pollution cost curve then slopes downwards.[25] The first point to notice is that a non-negative slope of the *total* pollution cost curve is not at issue. Pollution cost, for any particular pollutant, is expected to increase with the magnitude of pollution; at least until every conceivable damage produced by a pollutant has already been experienced. The question is whether each successive incremental unit of pollutant creates more pollution cost than the previous one: the debate therefore centres on the slope of the *marginal* pollution cost curve.

The conventional hypothesis that the slope is positive relies on two assumptions:
(1) that increments to the amounts of pollution have increasingly damaging effects measured in physical terms: an increase by one unit of the pollutant emitted by the polluter reduces the pollutee's output (or consumption) by more if the level of pollution is already high than if it is low. In other words, increasing marginal *physical* damage is assumed;
(2) that the *value* of a marginal unit of physical loss rises as the total amount of physical loss increases. This will be true if output yields diminishing marginal profit as it increases and consumers experience

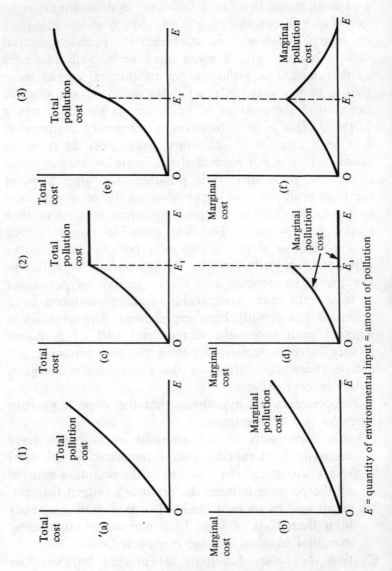

FIGURE 2.7   *Three cases of total and marginal pollution costs: (1) conventional formulation; (2) zero marginal damage case; (3) declining marginal damage case*

diminishing marginal utility from increased consumption. The marginal suffering (to the pollutee firm's managing director as measured by loss of profit, or to the consumer in the case of detrimental effects on consumption) would then become more acute as the total suffering mounted and the lower the remaining levels of undamaged production or consumption became.

The substance of the criticisms that have been made of the conventional hypothesis of a positively sloped marginal damage curve is easier to understand when it is realized that some writers concentrate on assumption (1) and some concentrate on (2).[26] The writers who challenge assumption (1) rely on some form of saturation thesis, in which it is postulated that the pollutee suffers zero or declining marginal physical damage at high levels of pollution. The following examples relate to the loss of output and profit by a polluted firm but similar reasoning, in terms of consumption and utility loss, can be applied to consumers.

In figure 2.7, quadrants (c) and (d), one possibility is illustrated: the pollutee suffers increasing marginal pollution cost as a result of increasing marginal physical damage (loss of output)[27] up to the amount of pollution E_1, but at that point the firm's costs are such that it is forced out of business and pollution increments thereafter create zero marginal damage. The resulting discontinuity in the marginal damage cost curve for one pollutee would not, however, be reflected in the aggregate curve, even if a number of firms were affected, since they would be unlikely all to go out of business at the same pollution level. If the number of firms going out of business gradually increased after E_1 the aggregate marginal pollution cost curve clearly could slope downwards even though each firm remaining in business experienced increasing marginal damage. This situation is shown in figure 2.7, quadrants (e) and (f).

From this brief summary it is apparent that the marginal damage cost curve slopes downwards only if the pollution has highly detrimental effects in the context of the pollutees' internal costs and sales revenue. One may be sceptical of the practical importance of the 'shutdown due to pollution' case, but it cannot be ruled out on theoretical grounds. It is not necessary, however, to rely on such an extreme effect as shutdown to generate a downward-sloping marginal pollution cost curve. If, in figure 2.7 (e) and (f), each successive increment of pollution above E_1 causes *less* extra physical damage (loss of output), then pollutees individually and in the aggregate will face a downward-sloping marginal cost of pollution curve segment.[28] This may be plausible for some types of pollution. For example, once the dumping of cyanide in a lake reaches a certain level of destructiveness extra units of poison may have a decreasingly detrimental *marginal* impact; similarly, a farmer's cereal crops may be initially highly sensitive to SO_2 but, after a point, as the level of pollutant concentration increases the marginal fall in crop yield may decrease until both the yield and marginal damage are zero. The crucial question is whether the $O-E_1$ or the E_1–upwards range of the marginal cost function in figure 2.7(e) is the *relevant* one for real-world pollution problems, a judgement about which the reader may care to speculate. What *is* clear is that the conventional upward-sloping marginal cost curve cannot be said to be established on theoretical grounds where diminishing marginal physical impact is imaginable.

The attack on the second assumption ((2) on p. 35) has been directed at the evaluation of the pollution damage sustained by consumers but not at that suffered by firms. Do consumers experience diminishing marginal utility from improvements in environmental quality, or in other words increasing marginal *dis*utility from the deterioration of environmental quality, as is conventionally assumed? Of course, if we assume that environmental destruction is available only in large lumps, so that the choice is between no destruction, almost complete destruction (saturation) by the first unit of pollution, and slightly greater total levels of destruction as pollution increases, then the stepped marginal damage (loss of utility) curve is *likely* to slope downwards! If a steel mill has been placed in what was a beautiful view, it should not surprise us that a second mill is less marginally damaging.[29] However if we step back and place that particular beautiful view in the context of numerous such views threatened by steel mills, it seems much more plausible that each successive view that is threatened with destruction will be more highly valued than the last. The loss of one remaining beauty spot in an area offering little in the way of beauty may be judged more serious than the loss of one beauty spot in an area abounding with natural beauty. Much depends on whether a view can be evaluated in isolation to the general character of the locality. If, for example, people derive pleasure from 'areas of natural beauty' rather than from individual beautiful views, then the loss from locating a mill in the beautiful area may have the more serious detrimental effects. It is very hard to generalize, but one proposition does seem justified: when the decision whether to build an extra factory is being made the general philosophy should be that the extra destruction of the environment in the aggregate will generate a greater loss in value terms than did the previous factory. As a scarce environment is used up the presumption should be that remaining units are more highly valued than those that have been ruined. Perhaps in certain industrially developed localities a point has been reached where, at least for *some* extra factories, the extra visual environmental damage will be slight, a fact that has implications for zoning policy (chapter 3), but this categorically does not imply a downward-sloping marginal damage curve in the aggregate.

In conclusion, on the question of the slope of the marginal damage cost curve, the attitude that will be adopted here is that the disputable nature of assumption (1) for *severe* pollution cases should not lead us to reject for general purposes an analysis based on upward-sloping marginal damage cost curves. However, it remains desirable to remind ourselves from time to time of the implications of extreme pollution, in particular for the workability of markets (see section 3.1.2 below).

Before considering the other characteristics of pollution costs it is appropriate to pause at this point and recognize that, quite apart from the slope of particular curves, the use of smooth abatement and pollution cost curves for analytical purposes abstracts from the complicated relationships between pollutants and the environment, and between different pollutants, that exist in reality. The total abatement cost curve would, for example, consist of a series of linear segments if firms were to abate by switches between a restricted number of production processes.[30] The important question is whether the use of idealized, regular curves invalidates the economist's theory as a representation of pollution control problems in the real world. The answer is that the shape of the curves, for example the slope of the marginal pollution cost curve, can have an important bearing on the character of the efficient solution to external costs (sections 3.1 and 3.2 below), but that it is not usually their smoothness that is critical. As long as the real-world curves allow an efficient solution to be identified the conventional theory can be applied.[31]

In a curious attack on the conventional economic analysis of pollution K. W. Kapp has argued that:

In the light of the foregoing discussion of the causal chain and complex interdependencies which give rise to a disruption of man's natural and social environment, it becomes evident that the conventional framework and tools of economic theory are ill-adapted and in fact irrelevant for the analysis of the phenomena under discussion.[32]

The 'foregoing discussion' referred to is a series of examples of pollution costs being determined by both the amount of pollution and by natural environmental variables (wind velocity, stream flow, etc.), and examples of pollutants interacting with each other to give rise to greater pollution costs than they would in isolation. Now the natural environmental variables affect the position of the pollution cost curve, but they are quite compatible with the existence of such a curve. Similarly, the interactions between pollutants would imply that for each pollutant a family of cost curves would exist one for each level of the other, interacting, pollutants, and that a single pollution cost curve could be drawn only for given amounts of the other pollutants. But again, the *idea* of a socially efficient amount of pollution retains its meaning; the only practical problem is that the efficient levels of different pollutants become interdependent.[33] The fact that causal chains *are* complex in the environment, as all biologists agree, and that interdependencies between pollutants undoubtedly occur means that estimating the position of pollution cost curves for real-world pollutants is a complex operation, but it does not mean that the economist's analytical framework, which indicates what we should like to be able to measure, is invalid or irrelevant. Nor incidentally does the fact that with economic growth the positions of the cost curves change prevent one from using such curves to analyse this dynamic situation. We can agree that the evaluation of pollution in the real world is more complicated than the smooth cost curves might suggest, yet still believe that external cost theory using such curves provides a powerful tool for the analysis of real-world pollution problems.[34]

The second main characteristic of aggregate pollution costs that will be assumed throughout is that they have an effect on the *marginal* decisions of those who bear them. This requires that pollution increases the marginal cost of affected firms or reduces the marginal utility or affected consumers. In other words, in the case of an affected producer, for example, an increase in the amount of pollution does not shift his total cost curve upwards parallel but rather increases the *slope* of the curve. This means that the impact of pollution cost is felt not only through an increase in the firm's fixed cost, which would leave his short-run output choice unaffected as long as the decline in profit did not drive him immediately out of business. In the long run of course all costs are variable anyway, so that any pollution must lead to a change in long-run marginal cost and influence the firm's decisions. The assumption that even short-run effects are felt at the margin will simplify the analysis, but it could also be argued that most real-world pollution cases do have marginal effects, for it is singularly difficult to think of ones where the only impact is on pollutees' fixed costs (such as an increase in the rent of the factory or the capital cost of existing plant).[35] In fact we shall assume that *only* the pollutee's variable costs are altered by pollution.

The third assumed characteristic of pollution costs is that they are unilaterally imposed by the polluters on pollutees; there is no reciprocal imposition of costs by the pollutees on the polluters. In other words the polluter and pollutee groups are mutually exclusive sets; no firm or consumer is both sinned against *and* sinning. That this is not always the case in reality is apparent to drivers in congested streets, smokers in crowded rooms and firms who have to clean polluted river water for production purposes and who then return it to the river polluted again. However, the insight gained from elaborating models of reciprocal external costs is not sufficient to justify the considerable increase in analytical complexity, and attention will be limited to the widely applicable unilateral case.[36]

The final characteristic of pollution costs concerns the size of the group of pollutees and the way in which the costs are spread across the group. This characteristic will prove to be significant in the examination of the reasons why the market system fails to solve the problem of external pollution costs (section 3.2 below) and will have important implications for the efficiency of attempts to solve the problem through a tax subsidy scheme (chapter 4). In the theory of public finance it has proved fruitful to distinguish between private goods and public goods, and the distinction will be directly applicable to pollution costs.[37] *Private* goods, such as food, are those that yield satisfaction only to those to whom they are supplied (other people being excluded from participation), and that can be shared out among the consumers in society, subject to the limitation that an increase in one person's consumption can be achieved only by an equivalent decrease in the quantity of the good consumed by someone else. *Public* goods, such as defence and flood control, on the other hand, are jointly supplied to consumers in such a way that once they are supplied to one person they become equally available to all. While person A's own consumption of the good gives no satisfaction to B, the *supply* of the good to A does yield satisfaction to B. This means that supplying the good to person A does not require an equivalent reduction in the quantity of the good consumed by B or anyone else. Clearly the supply of the good to A produces a benefit to B that will be an external benefit if A cannot exclude B from using the good without payment.

It does not require much thought to realize that the

equal-availability characteristic (perfect jointness in supply) is rarely found in reality. Many goods have an element of jointness, but the availability varies between groups of the population (the police service may offer greater protection in some districts of a city than in others; a bridge is more available for use to those who live in the locality than to those further away; defence systems may concentrate their protective capability in the region where the government is located). Many goods also display (external) benefits to people other than those buying them, so can we conclude that the two properties of jointness in supply and external benefits are inextricably linked together? Head has shown quite clearly that the answer in no.[38] In the first place, while it is true that many goods (i.e. defence, public health programmes) that display jointness in supply also provide problems of excluding non-purchasers (yield external benefits) and fall neatly into the category of public goods, there are also goods that require joint supply but for which exclusion is technically and economically possible (e.g. sporting events, various forms of transport). Jointness does not necessarily imply non-exclusion; moreover, difficulties of exclusion do not necessarily imply jointness. In the case of natural resources, for example, often it is impossible for the firm that initiates extraction to exclude other firms from participating (reaping external benefits from the pioneering firm's exploration), but unlike jointly supplied goods the increase in extraction by one firm *does* reduce the availability of the resource to other firms. While the two characteristics of public goods are distinct, undoubtedly the coincidence of the two in particular goods does strengthen the case for public intervention owing to the under-supply of the goods by the market, intervention that might take the form of subsidizing or publicly organizing the supply of the goods to consumers. While many, perhaps most, goods in the real world fall into the class of intermediate goods, displaying in varying degrees the features of private and public goods, it will be convenient for our application of these ideas to think of goods being dichotomized into pure public goods, which display both characteristics, and pure private goods, which display neither.

Let us return from this foray into the public financiers' domain to consider the spread of pollution costs across pollutees. The generation of pollution costs by firms or consumers is the converse of the production of *goods*. Consequently we may think of pollution costs as being either 'private bads' or 'public bads' depending on whether they have the converse of the characteristics of private or public goods. A pure private bad is one that is imposed exclusively on one pollutee and is subject to the limitation that a decrease in the pollutee's 'consumption' of it can be obtained only at the cost of an increase in another pollutee's 'consumption'. Similarly, an increase in one pollutee's willingness to accept the pollutant reduces the quantity others must accept. Thus, if a farmer who lives near a large city makes extra space available for land-fill with solid refuse, then less will have to be disposed of in other places. The dumping of extra refuse on a farmer's land does not create a cost that is equally imposed on all, although such dumping would clearly be more of a public bad if there were general access to the area as in a public park. It seems, however, that private bads are not common.[39] The majority of air- and water-borne pollutants have a distinct element of publicness. In the extreme the perfectly joint supply of a bad to all pollutees means that an increase in person A's exposure to the pollutant does not reduce person B's exposure; on the contrary they increase together. The fact that the DDT content of one person's body fat has risen does not reduce anyone else's; the effect of nuclear generators and explosions on background radioactivity levels is quite undiscriminating (although local variations in intensity do occur) and people do not have to compete for their normal dose. Many pollutants have locally high concentrations, near to sources of emission, but the more persistent the pollutant the more nearly pure is the public bad, owing to the effectiveness of the transportation of residuals by air and water.

In addition to being equally available, at least within a locality, a public bad is non-rejectable. If more of the bad is supplied to person A then person B cannot avoid suffering as a result.[40] This characteristic is the counterpart of non-excludability in the public good case. Abating the pollution by altering the polluter's technology would yield a public good in the form of reduced pollution cost; and to abate in this way for A would provide B with a benefit from which he could not be excluded.

In essence therefore a public bad is an external cost that is jointly supplied to a group of pollutees, and the amount of pollution received by any individual member of the group has no effect on the amount received by others. Since this seems to be the common characteristic of most real-world pollutants, much of the subsequent analysis will assume that pollution is a public bad. It need not, however, be assumed that the group of pollutees is necessarily very large. A pollutant may be supplied to a single pollutee in a particular locality, yet retain the characteristics of a public bad, so that anyone who moves to that locality will be unable to reject the (joint) supply of the pollutant. It will be convenient at one of two points in the next chapter to distinguish cases of large and small numbers of polluters and pollutees, but the pollutant will still be regarded as inherently public. For the publicness of pollution is an important contributor to the failure of markets.

2.4 PROPERTY RIGHTS: A PRELIMINARY VIEW

When the correct interpretation of the term 'external cost' was discussed in section 2.2 it was mentioned that pollution costs may lie outside the calculus of the polluter if he is allowed by law to ignore them. Before considering further the sources of market failure to internalize such costs (section 3.2 below) we should clarify the economist's use of the term 'property rights' in the context of market solutions to external cost problems.[41]

One of the important functions of the legal system is to define the legitimate uses of property. Even in a social system based on the private ownership of property, society, through the legal system, confers on the owners only limited rights of use. A land-owner, for example, may not build on his property buildings that are of unsafe design; he may not store high explosives or harbour dangerous wild animals; nor may he obstruct public rights of way. Man-made possessions are similarly set about with restrictions as in the cases of guns, cars and other inherently dangerous chattels. It is clear, therefore, that the ownership of property is essentially the ownership of the rights to some, but not all, of the services that property is capable of yielding. In addition to allowing the owner to enjoy these services, private ownership normally provides for the exclusion of others from also enjoying the permitted activities, although even with property narrowly defined as land this exclusion is subject to a qualification where rights of access and rights of way exist. The other main right conferred by private ownership is the right to sell the title to the property, but again not without restrictions. One may sell a second-hand car, but only if it is in roadworthy condition; one may sell land, but not necessarily the right to alter the use of the land say from agricultural to house building use. However, it is possession of the permitted services of property (including the service of appreciating in value!) that induces people to pay prices for all forms of goods and natural resources. Market prices are, therefore, the prices of property rights: goods that offer no services (yield no utility) will not be demanded and will have a zero market price.

A polluter is likely to ignore the consequences of his activity for others if the (potential) pollutees clearly do not have a right to freedom from the interference, if the rights of pollutees are ill defined, or if the pollutees nominally possess the right to freedom from interference but the right is not enforced. This follows from the fact that the expected cost of not polluting (abating) is generally positive and the expected cost of polluting to the polluter is the multiple of the probability of being forced to pay and the amount to be paid (equal to the estimated pollution costs with or without a fine added). If the probability of being detected and obliged to pay approaches zero then the cost-minimizing choice is obviously to pollute. As the probability rises the behaviour that minimizes expected cost depends on the relative sizes of the abatement cost, pollution cost and fine.[42]

But why should pollutee rights be non-existent, ill defined or unenforced? A thorough explanation would require an investigation of legal history but one or two general points can be made.[43] The definition of property embodied in common law is generally rather narrow, placing most emphasis on the defence of the physical or tangible features of the property, land in particular. This narrowness of definition dates from a time when the law was mainly concerned with the protection of the property-owning class from the loss of value of land owing to physical damage, loss of vegetation and livestock, resulting from the intrusion of neighbouring activities and deliberate acts of criminal interference (poaching etc.). Since that time the pressure of increasing population and output, more sophisticated technology and more complex products (culminating in the non-organic, persistent chemicals rapidly developed over a short time period) has meant that the attack on the services derived from property is now less tangible, and derives from a more complex, and less local and identifiable, set of sources. Thus, the owner of a tree nursery may be able legally to protect himself from the intrusion of his neighbour's cattle but be helpless to prevent the growth of his trees deteriorating as air pollutants are carried from a distant town. The law may stop your neighbour from burning a bonfire twenty-four hours a day but leave you without protection from the all-pervasive odour from a nearby chemical factory. By and large it is the air and water associated with land that is threatened, but the continually changing character of pollution leaves a static common law structure lagging behind. The judges may adjust gradually to recognizing the importance of the sulphur dioxide content of the air, but how long before the lead, mercury and cadmium content is identified and made the subject of litigation?

However, the sluggishness of the common law of private property is not the only, or perhaps even the main, cause of gaps in the range of effective pollutee property rights. Many of the most serious pollution problems relate to property that, nominally at least, is owned in common by all of us. In developed capitalist societies substantial assets are owned by the state; parks, roads, rivers and many lakes, forests and beaches, not to mention the atmosphere about us and the soil beneath, all are 'common resources'.[44] By their nature common resource rights may effectively be enforcable only by government, yet until recent years largely unrestricted access to such resources was the general rule. When resources seem almost limitless in supply there is no apparent benefit from restricting access, no gain from establishing a positive price for their use. The clamour for pricing or some other form of restricting access to an area of natural beauty occurs only when congestion begins to reduce the quality of the area. Similarly, when the monitoring of river and air pollution levels indicates low concentrations of significant pollutants the demand for restricting the access of polluters is weak. The inherent danger with common resources is that in the absence of an enforcement agency with carefully defined responsibilities in maintaining their quality, common resource property rights lead to unlimited use and to eventual conflicts between different types of use. Unrestricted access to common resources leads to over-exploitation, whether the resource is wildlife (buffalo, whales, kangaroos), clean air and water, or even timber, underground minerals and other materials. The trend in many countries is towards the extension and tightening of statutory controls over the utilization of common resources, and in chapter 5 we shall consider some of the issues relating to the application of such controls to pollution.

CHAPTER 3

Market Failure

In the previous chapter the main components of the theory of external costs were explained. An understanding of these components, although not in itself offering intellectual delights, is necessary for analysing the limits to markets and the characteristics of the available forms of government intervention. Assembling the components of the theory provides a framework to help us identify the consequences of external pollution costs for the efficiency with which resources are allocated in the economy. This will be the main task of section 3.1, although the discussion at some points will be broadened to include the implications of pollution for justice, a consideration that is noticeably lacking in many economic studies of pollution. Section 3.2 presents a summary of the difficulties associated with solving pollution problems through the operation of markets. Initially the character of market solutions in an idealized form is described; and then we come closer to reality and examine the various obstacles that may obstruct the achievement of socially efficient outcomes without government intervention. The pessimistic conclusion concerning the efficacy of markets provides the setting for the evaluation, in chapter 4, of a variety of possible instruments for pollution control.

3.1   External Costs, Justice and Efficiency

3.1.1.   *External costs and justice*

The existence of external pollution costs has two possible effects on the way the economy operates to generate benefits, or welfare, for society. In the first place, in the absence of any private or government compensation arrangements, pollution leads to uncompensated losses being incurred by pollutees which may be thought unjust. Second, external pollution costs are likely to move the allocation of resources between alternative uses away from the allocation that would prevail in their absence. If this change in resource allocation is detrimental, as it is likely to be, the situation with pollution represents a *mis*allocation, meaning an inefficient allocation of resources. These issues of justice and allocative efficiency will be discussed in some detail, but first a cautionary tale concerning efficiency analysis and economic policy.

The writing of many economists and lawyers on the problems of pollution control have concentrated almost exclusively on the *efficiency* of alternative solutions.[1] A brief homily on the limitations of economic theory is therefore in order: simply stated, there is nothing in economic theory to suggest that a particular outcome of markets, or of policy, that is economically efficient, in the sense of maximizing the *sum* of net benefits to society, will necessarily be preferred by society as a whole. An efficient structure of the economy may well have undesirable consequences which society would prefer to be weighed in the balance in the decision-making process. This possibility becomes clearer when we consider further what is meant by 'a change that maximizes the net gains to society'. Net gains are to be interpreted as the excess of gainers' gains over losers' losses. If there are no losers and some gainers the move is said to be Pareto-efficient; if there are some losers and some gainers, but the sum of losses is outweighed by the sum of gains, the move is said to be potentially Pareto-efficient.[2] Many analyses of pollution control adopt a definition of efficiency that encompasses both of these possibilities, and proceed to conduct a search only for the organization of the economic system (markets or policy instruments) that yields the highest net gains to society.[3]

The problem with relying on an efficiency analysis of the alternative solutions of pollution problems is that the source of the 'problems', and therefore the origin of pollution control policy, lies in the need to resolve or mitigate conflicts of interest. Consequently a change in the legal rules (e.g. a change in property rights in favour of pollutees) or the introduction of a pollution control policy is likely to make someone worse off than before. To judge whether decisions that yield net gains to society, i.e. that are *efficient*, are also socially *optimal* (i.e. are ideal given all society's objectives), consideration must be given to the consequences for the losers. Welfare economists have generally regarded this as a question of judging the desirability of the effect of the move on the distribution of income (or welfare), and have argued that the move will be socially optimal as long as the net gains are not accompanied by an 'undesirable' redistribution of income from losers to gainers. Judgements on the desirability of a change in the distribution of income require some notion of distributive equity (for example, that it is fair for income to be more equally distributed). But in the context of the external costs imposed by one group on another, the implications of say a change in property rights for the *general* distribution of income in society may be less important than the *particular* income (and other welfare) consequences for the gainers and losers from the change. In fact, the legal system or the pollution

curves that slope downwards?[6]
(2) Does the analysis of socially efficient resource alloca-
     tion in the presence of external costs need to be
     modified if there is a large number of polluters and/or
     pollutees?
(3) How can the analysis take account of flexible tech-
     nology and a choice of location for the polluter?[7]
(4) If external costs could be internalized costlessly
     would a movement towards a socially efficient out-
     come be generated if the goods market for which
     the polluter produces is monopolistic?

Imagine a firm, A, which is operating in competitive
goods and factor markets and whose fixed technology,
that is its single available process, generates pollution
of the 'public bad' variety (see chapter 2, pp. 45–6)
at a fixed ratio to output. This pollution raises the costs
of production of another firm, B, which is in a different
industry, but which is also operating in competitive goods
and factor markets. The result of pollution by firm A
is to make the production costs of the pollutee, firm
B, dependent both on its own output, q^B, and on the
polluter's output, q^A. The pollutee B faces a rising marginal
cost curve even if pollution is zero (that is if firm A
does not produce), shown by $MFC^B(q_0^A)$ in figure 3.1 (b).
But as A's output and pollution increases, B's marginal
cost curve pivots upwards. At any level of B's output
the effect on B's marginal cost of a unit increase in
A's output is greater the higher is A's output level.[8]
This means that increases in A's output impose an *increasing marginal external cost* on B in the form of lost profit.
This is represented diagrammatically by drawing the
MFC^B curves so that they pivot upwards, in response
to a unit increase in A's output, less when A's output
is low (such as q_0^A) than when it is high (such as q_3^A).
For any particular level of the pollutee's own output,
such as q_2^B, the marginal external cost of a unit increase

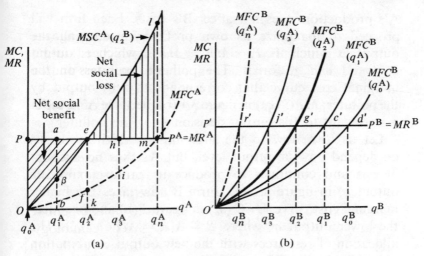

FIGURE 3.1 *The effect of pollution on resource allocation: (a) polluter A's output choice; (b) pollutee B's output choice*

in A's output is the vertical distance between the MFC^B curve before the increase and the MFC^B curve after it. Adding this marginal external cost to the polluter's own marginal private cost curve, MFC^A in figure 3.1 (a), yields the marginal *social* cost curve $MSC^A(q_2^B)$ for the polluter's output. Since the magnitude of the external cost depends on the output levels of both A and B, there will be a family of marginal social cost curves in figure 3.1 (a), one curve for each level of B's output. Each marginal social cost curve is the extra cost to society (that is to A and B together), of A's production if B were to choose a particular level of its output q^B.

The information that has just been described is enough to enable us to demonstrate the effects of external pollution cost on the allocation of resources between the two competing uses in this simple society, namely the production of A's good and the production of B's. Consider first, as a basis for comparison, the situation in which

A's production does not affect B's costs. Each firm will proceed to maximize its own profit by producing the output at which its $P = MR = MFC$, which is output q_n^A *for A and* q_0^B for B. The pollutee operates on the marginal cost curve that corresponds to zero output by the polluter, $MFC^B(q_0^A)$ in figure 3.1 (b), because A's output is irrelevant to B's output decision.

Let us turn now to the situation in which B's costs do depend on A's output level, but A takes no account of this and continues to produce its profit-maximizing output q_n^A in figure 3.1 (a). Firm B now faces the higher marginal cost curve $MFC(q_n^A)$ in figure 3 (b) and produces the lower output q_n^B where $P = MR = MFC^B(q_n^A)$. The allocation of resources with the new output combination q_n^A, q_n^B is clearly different from the one that obtains in the absence of the cost interdependence, q_n^A, q_0^B, since $(q_0^B/q_n^A) > (q_n^B/q_n^A)$.

This change in the allocation of resources can be shown to lead to a resource *mis*allocation, a move to a less efficient position, by comparing the output combination q_n^A, q_n^B with the combination that yields the highest net gains to society. The problem with the combination q_n^A, q_n^B is that firm A is producing some units of output for which the marginal cost *to society* exceeds the valuation that consumers place on them (as represented by the price they are prepared to pay for them) and that therefore create a net loss to society. Examining the gains and losses at a series of possible output levels for firm A should make this clear. If A were initially producing nothing (q_0^A) and B were maximizing profit at q_0^B, consider the consequences of A then producing output q_1^A instead. The profit gained by A from this move would be *Opab* in figure 3.1 (a), the excess of marginal revenue over marginal cost (MFC^A); but B would suffer a loss of profit equal to $O'c'd'$ in figure 3.1 (b), the area between $MFC^B(q_1^A)$ and $MFC^B(q_0^A)$, as it reduced output to q_1^B. Assuming

$Opab > O'c'd'$ then the move to q_1^A yields net gains to society, i.e. it is socially efficient.[9] A move on from q_1^A, q_1^B to q_2^A, q_2^B is similarly efficient *for society* if $baef > O'g'c'$. But if $fehi < O'j'g'$ then a further increase in A's output to q_3^A, accompanied by B's reduction in output to q_3^B, is *not* socially efficient.[10] The output configuration q_2^A, q_2^B maximizes the sum of A's and B's profits and is said to be the socially efficient situation.

The marginal social cost of the q_2^A-th unit of A's output is ke in figure 3.1 (a), which is equal to the price, P^A, of A's output. The marginal social cost ke is the sum of the marginal private cost kf and the marginal external (pollution) cost fe. Clearly, fe must be the *socially efficient* level of marginal pollution cost. The marginal social cost ke is a point on the marginal social cost curve $MSC^A(q_2^B)$ which is drawn on the assumption that the pollutee B's output is at its socially efficient level q_2^B. For A's outputs above q_2^A it is clear that $MSC^A(q_2^B)$ exceeds the consumers' evaluation of the product represented by P^A. Consequently the extension of output from q_2^A to q_n^A, which A prefers as its profit-maximizing output, imposes the social loss (of profit) elm, in figure 3.1 (a). This social loss is the reduction of joint (A plus B) profits below the maximum attainable at q_2^A, q_2^B where $P^A = MSC^A(q_2^B)$.[11] The condition that A's price and marginal social cost are equal is sufficient to guarantee social efficiency, given that B's output is at its socially efficient level, q_2^B.[12]

It is important to note that allowing firm A to pollute at will not only leads to a level of A's output in excess of the socially efficient one, but also causes B to reduce its output to q_n^B, which is below the socially efficient level q_2^B. The achievement of social efficiency does not, however, imply that B will revert to the output level q_0^B, which it would choose if it were unaffected by A's production activity! The reason for this is that, once the interdependence of the firms' costs is recognized, it becomes apparent that an increase in either firm's output increases the social costs created by the polluter's activity. An increase in B's output raises the damage resulting from, and therefore the social cost of, A's output. Another way of looking at this reciprocal relationship between the two activities is to observe that B's production can be protected from interference only by a sacrifice of output, and therefore of profit, by firm A.[13] Only if such protection were costless because A's output (at the margin) were unprofitable could B's output q_0^B be socially efficient.

The socially efficient levels of q^A and q^B have been found by comparing, for each level of q^A, the marginal benefit (profit) to A of an increase in its output and the associated marginal cost (loss of profit) to B. The marginal benefit of producing and polluting is of course also the marginal cost of abating pollution under a rigid technology (see section 2.3 above). The marginal cost of pollution incurred by B is the marginal benefit of pollution abatement.[14] It should come as no surprise therefore that the characteristics of the socially efficient situation can also be described in terms of marginal abatement cost and marginal pollution cost curves. Let us convert the previous analysis of the socially efficient state into this form.

In figure 3.2 (a) the curve MAC^A plots the marginal benefit (profit) that the polluter A receives from its different output levels, that is the marginal abatement cost. At output q_1^A, for example, the cost is *ab* in figure 3.1(a), the marginal profit represented by the vertical distance between MR^A and A's marginal private cost MFC^A. This marginal abatement cost is now plotted as *us* in figure 3.2 (a).

At output q_1^A the marginal pollution cost (loss of profit) incurred by B, *if B is producing at its socially efficient level* q_2^B, is the difference between $MSC\ (q_2^B)$ and MFC^A in figure 3.1 (a), that is $b\beta$, which can be plotted as

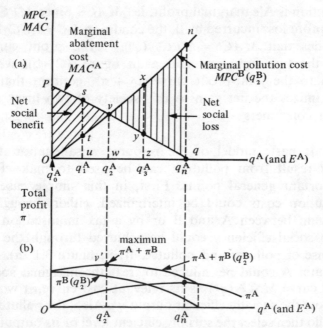

FIGURE 3.2 *Pollution and resource allocation: an alternative description*

ut on the marginal pollution cost curve MPC (q_2^B) in figure 3.2 (a). Clearly, up to q_2^A the cost of abating pollution at the margin exceeds the benefit from abatement, so that output increases up to that level increase the profits of A and B together (see figure 3.2 (b)) and therefore are socially efficient. Output rises above q_2^A are not socially efficient. The condition for social efficiency in figure 3.2 (a) is $MAC^A = MPC^B$ (q_2^B); i.e., marginal abatement cost equals marginal pollution cost. This is merely another way of stating the condition $P^A = MSC^A$ (q_2^B) in figure 3.1 (a). The condition $P^A = MSC^A$ (for output q_2^B) can be written as $P^A = MFC^A + MPC^B$, since social cost is the sum of A's production cost MFC^A and B's loss of profit owing to pollution (MPC^B). This statement implies $P^A - MFC^A = MPC^B$, where the left-hand side of the equation is A's marginal profit, i.e. MAC^A. Since MPC^B is the profit loss incurred by B, the condition $P^A = MSC^A(q_2^B)$ implies that $MAC^A = MPC^B(q_2^B)$.[15] The profit curves for the two firms, π^A and π^B in figure 3.2 (b), which sum to the joint-profit curve $\pi^A + \pi^B$, confirm that q_2^A maximizes the net gain to this society of two firms and their consumers.

This basic model of the resource misallocation that may result from pollution can be used to make four important general points. First, in this simple case, if pollution costs could be internalized, either through a bargain between A and B or by a tax imposed on A, then social efficiency could be achieved through the response of polluter and pollutee. If, in figure 3.1 (a), the polluter A could be induced to treat the marginal social cost curve $MSC^A(q_2^B)$ as its relevant cost curve, it would choose the socially efficient output q_2^A. Also, the pollutee B would then select the socially efficient level of *its* output q_2^B.

Second, the conditions that guarantee social efficiency in the conventional case, namely that $P^A = MSC^A(q_2^B)$ and $P^B = MFC^B(q_2^A)$, do not involve the payment of compensation to the pollutee, B. Yet at the socially efficient point g' in figure 3.1 (b) the pollutee, firm B, is suffering a loss of profit $O'g'd'$. We shall return to this question later (section 4.2 below), but let us note at this point the consequence for social efficiency of paying equivalent compensation in a particular form. Consider the possibility of a tariff of compensation set equal to marginal pollution cost at each combination of q^A and q^B. This tariff may be thought of as a price paid to firm B for accepting a particular level of pollutant analogous to the price paid by a consumer for receiving a good. The purpose, in efficiency terms, of attaching a price to a good is to force the consumer to take account of the benefit from the good foregone by other people. For a private bad the price paid to B for accepting it would encourage B to take account of the benefit to others of having a reduced exposure to the bad, and thereby would encourage B to take more of the pollutant as long as the damage it suffers at the margin does not exceed the price (compensation) received.[16] But with a public bad of the kind we have been assuming there is no benefit to other pollutees if B accepts more of the pollutant that is jointly supplied to all. There is no advantage in encouraging firm B to accept more of the pollutant, *except* if its doing so leads to a social gain by allowing A to produce more. We have seen, however, that A and B will automatically adjust efficiently as long as A faces the social cost of its activity. Offering compensation to B does not assist the socially efficient balancing of their two activities. The very existence of the pollution that results from A's socially efficient level of output will provide firm B with the required incentive to adjust to the socially efficient output q_2^B.[17] If the compensation tariff is not *needed* for a socially efficient response by A and B, would it nevertheless be compatible with such a response? If firm B were offered equivalent compensation $O'g'd'$ for the damage resulting from A's socially efficient output q_2^A, then it would become indifferent between its outputs q_0^B and q_2^B. Both of these outputs would yield the profit $O'p'd'$ in figure 3.1 (b). Firm B would then have no incentive to reduce its output below q_0^B. With external costs internalized, firm A would face a marginal social cost curve MSC^A (q_0^B) (not shown) to the left of MSC^A (q_2^B) in figure 3.1 (a), and the socially efficient outputs would not necessarily be produced.[18]

The third point derived from the basic model is that with a fixed technology and a strong preference of the polluter for the pollutees' locality, pollution control necessarily involves a loss of output (for example, q_n^A to q_2^A in figure 3.1) and probably, therefore, a loss of employment in the polluter's activity. On the other hand there will be an increase in B's output (for example, q_n^B to q_2^B in figure 3.1) and employment, but resistance to pollution control policy in practice often centres on the impact on the polluter's activity. It is important to consider the seriousness of this problem if the technology (or location) is flexible, which we shall do later (p. 70).

Fourth, we should note that the socially efficient level of polluting activity is identified on the assumption that the marginal pollution cost curve is in a particular position. Yet it has been suggested that the very act of polluting at an (initially) socially efficient pollution level that exceeds the assimilative capacity of the environment might *reduce* the future assimilative capacity.[19] If this were so the declining assimilative capacity would cause the marginal pollution cost curve to pivot upwards, with the result that the socially efficient pollution level falls. This, it is suggested, will continue to occur unless pollution is kept within the assimilative capacity. Whether this predicted decline in the socially efficient pollution level is an empirically relevant problem in the real world depends critically on the extent to which assimilative capacity is reduced by pollution. The evidence quoted often relates to very high levels of water pollution (such as in Lake Erie), and it is far from clear that a system operating at an (initially) socially efficient pollution level would suffer severely declining assimilative capacity. In fact, the policies that are discussed in chapter 4, by reducing pollution levels, might improve the assimilative capacity and thereby retard any long-run decline in its level that has occurred in the state of uninhibited polluting.

Having presented the simple case of resource misallocation due to pollution, we now turn to the four questions posed earlier (pp. 55–6).

(1) *A downward-sloping marginal pollution cost curve*

In section 2.3.2 we discussed the possibility that, in cases of severe pollution, the marginal pollution cost curve might slope downwards, at least over some of its length. The conclusion reached was that this possibility could not be ruled out on theoretical grounds. The efficiency analysis can easily accommodate a downward-sloping marginal pollution cost curve; in fact, it does not require very much in the way of new analysis to show that the socially efficient solution is similar in principle to that in the conventional model. However, when it comes to examining the operation of market solutions, and of tax solutions in the absence of perfect information (section 4.3.2 (1) below), we shall have to consider the potentially powerful argument that the downward-sloping marginal pollution cost curve raises serious difficulties for both markets and taxes that aim at socially efficient solutions.

In figure 3.3 two possible cases of downward-sloping marginal pollution cost curves are drawn, case 1 in quadrants (a), (b) and (c) and case 2 in (d), (e) and (f). The marginal social cost curves in (a) and (d) are drawn for the socially efficient level of B's output, q^B_1 in case 1, q^B_0 in case 2.[20] The same is true for the marginal pollution cost curves MPC^B in quadrants (b) and (e). It is clear that in each case there are two levels of A's output at which price equals marginal social cost, q^A_k and q^A_1 in case 1, q^A_r and q^A_s in case 2.

In case 1 there is an initial marginal social loss from the expansion of A's output from zero, because marginal pollution cost exceeds marginal abatement cost, i.e. $MPC^B > MAC^A$. Consequently the total joint profit curve in (c) slopes downwards to a minimum at q^A_k, where $P^A = MSC^A (q^B)$. The equality of price and marginal social cost here identifies the socially *in*efficient point k,k' at which social gains are *minimized*! As A's output increases

NSB = net social benefit
NSL = net social loss

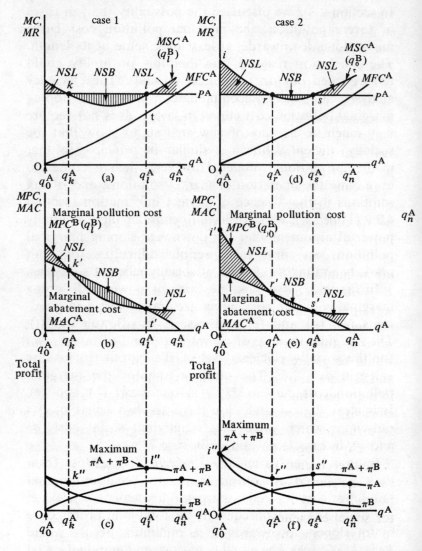

FIGURE 3.3   *Social efficiency with a downward-sloping marginal pollution cost curve*

further we find that $MAC^A > MPC^B$, so that joint profit increases to a *maximum* at q_l^A, where $P^A = MSC^A (q_l^B)$ once more. As the curves are drawn in quadrants (a) and (b) the social gain from output q_k^A to q_l^A exceeds the social loss up to q_k^A, so that total joint profit at q_l^A exceeds that at q_0^A. Therefore q_l^A is the socially efficient level of A's output, but we have found that $P = MSC^A (q_l^B)$ is not a sufficient condition for social efficiency if the marginal pollution cost curve slopes downwards. The sufficient conditions in case 1 are that $P^A = MSC^A (q_l^B)$ *and* that for a small change in q^A total joint profit falls, as it does at q_l^A.[21]

In case 2 the net social loss resulting from firm A's outputs from zero to q_r^A exceed the social gain produced by its outputs q_r^A to q_s^A. Although $P^A = MSC^A (q_0^B)$ at both q_r^A and q_s^A, neither output is socially efficient. At q_r^A the situation is similar to q_k^A in case (1): joint profits are minimized. At q_s^A on the other hand the condition $P^A = MSC^A (q_0^B)$ does locate a maximum point on the total joint profit curve, s'' in (f), but because the total net social loss up to q_r^A exceeds the gain from outputs q_r^A to q_s^A this point is only a *local* maximum. Joint profit would be greatest if A stopped producing in B's locality and B were allowed to produce unhindered, because q_0^A is the *global* maximum. This corner solution is socially efficient, yet price is not equal to marginal social cost![22] It turns out, therefore, that if the marginal pollution cost curve slopes downwards the condition $P^A = MSC^A$ is neither necessary (in the event of a corner solution such as i'' in (f)) nor sufficient (since it applies at minimum points, k'' and r'', and also at local maxima such as s'') for social efficiency.

What are the main implications of this efficiency analysis? In the first place, it means that a glib rule that a polluter's price be set equal to marginal social cost will lead to social efficiency only if the marginal pollution costs increase with the level of pollution and if the extreme solutions (namely, no output by the polluter or unrestrained polluting) are inefficient. Whether these conditions prevail can only be settled empirically: ideally we need to know the positions and the slopes of both the marginal abatement and marginal pollution cost curves. In many cases this information may be hard to obtain and the only way ahead is a sub-optimizing approach with pollution control policy aiming at pre-determined targets (see section 4.3.2.(2) below).

Second, despite the complication of multiple points of equality between price and marginal social cost, if the policy-maker *did* have this kind of information a polluter tax could be devised that would guarantee a socially efficient outcome.[23] We have seen that this is so in the conventional model (see p. 62), but it can also be shown to be true even if the marginal pollution cost curve slopes downwards. In case 1 in figure 3.3, if the pollution control agency had sufficient information to identify the output q_l^A, q_l^B it could induce the firms to move to this position by setting a tax on A's output of $lt = l't'$ per unit. This would place A on a marginal cost curve parallel to MFC^A passing through point l in quadrant (a), and the firm would choose q_l^A.[24] Firm B would respond by producing q_l^B. Notice that with this tax rate firm A will not select output q_k^A because the firm's price exceeds marginal factor cost plus tax right up to output q_l^A. Similarly, in case 2 a tax equal to Oi would place firm A on a marginal cost curve parallel to MFC^A and starting at point i, in quadrant (d).[25] No output would be profitable for A in B's locality so that q_0^A, q_0^B, the socially efficient output combination, would result from the tax.

(2) *Large numbers of polluters and/or pollutees*

It will be remembered that we have limited our attention to pollution that has the public bad characteristic that an increase in the level of pollution received by one pollutee does not alleviate the suffering of other pollutees. In many instances such pollution is created by many polluters, or it affects large numbers of firms or people, so that we should consider to what extent our analysis needs to be modified to take account of large numbers. Again for simplicity we use the example of polluting *firms*, but the costs and benefits could equally well be the losses or gains of utility by people.

The existence of a large group of polluters or pollutees raises serious problems for the workability of bargaining solutions (see section 3.2.2 below), but the previous analysis of social efficiency needs little alteration. All that is required is to interpret the MSC^A curve as representing the social cost of the output of the polluter *group* (firms $A(1), A(2) \ldots A(n)$) for any level of output of the pollutee *group* (firms $B(1), B(2) \ldots B(n)$). In figures 3.1, 3.2 and 3.3 the q^A and q^B axes now measure the aggregate of output for the relevant group; and the vertical axes measure the aggregate of costs or benefits for the group.[26] It follows of course that the socially efficient solutions must now be re-interpreted. Thus the efficient outcome q_2^A, q_2^B in the conventional model refers to the outputs for the two groups and if, for example, all of the polluters face the same cost curves then the q_2^A aggregate output will be equally shared between them.

Finally, for much of the analysis in subsequent chapters we shall identify the polluters' individual marginal pollution cost curves. Each curve is the sum of the pollution cost incurred by individual pollutees as a result of the particular polluter's emissions. Because pollution is a public bad each pollutee faces the same pollution level. Any polluter's marginal pollution cost curve is therefore the vertical sum of the marginal pollution costs incurred by the pollutee group at each level of pollution by that polluter.

(3) *Flexible technology and location*

The efficiency theory has so far assumed that the polluter's output and pollution levels are inextricably linked, but we know from the explanation of abatement cost in section 2.3 that a change in production process can alter the amount of pollution generated by a given output level. Figure 3.4 presents the efficiency analysis (using a conven-

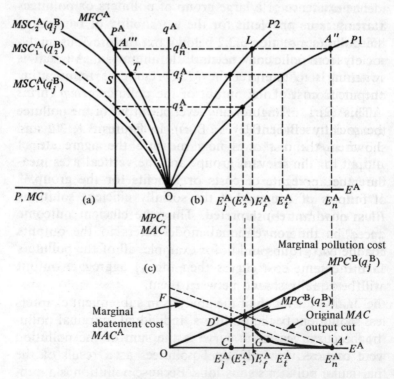

FIGURE 3.4 *Efficiency with a flexible technology*

tional upward-sloping marginal pollution cost curve) modified to allow for process-switching by the polluter. In quadrant (c) we have the marginal pollution cost curve MPC^B (q_j^B), and the polluter's marginal abatement cost curve $FD'GA'$, taken from figure 2.3 (b) in chapter 2.[27] It will be remembered that $FD'GA'$ is the lower segments of the intersecting MAC process switch and the new, lower, MAC output cut curve resulting from the initial process-switching. In figure 3.4, quadrant (c), the original fixed technology MAC output cut and marginal pollution cost curves are inserted for comparison.[28]

As the new curves are drawn the intersection D', at which marginal pollution and abatement costs are equal, identifies the socially efficient level of A's pollution, E_j^A. Starting from the pollution level E_n^A at point A', where the polluter pollutes unhindered, there are net gains to society from pollution abatement initially through process-switching to E_f^A, and then by means of a reduction in output, from E_f^A to E_j^A.

This part of figure 3.4 does not, however, tell us the socially efficient level of *output* for firm A: this is shown in the rest of the figure. At the starting point, A', A'', A''' in the three quadrants, the polluter is using the most polluting process $P1$[29] to produce the level of output q_n^A that maximizes its own profits. The curve $P1$ in quadrant (b) shows the level of pollution (E^A) produced by different levels of A's output if process $P1$ is adhered to: the point A'' shows that output q_n^A pollutes to the extent E_n^A.[30] Once the polluter abates from E_n^A at A' to E_f^A by switching process (and moving along the MAC^A curve) the new, less polluting process implies less pollution per unit of output so that in quadrant (b) the curve shifts from $P1$ to $P2$. If process switching were costless it would be possible to keep on shifting this curve further as pollution fell (eventually to zero on a path that included points such as L) with output unchanged. This would mean that the marginal social cost curve would pivot downwards from MSC_1^A (q_j^B) in quadrant (a) to a position coincident with MFC^A: the external cost would be eliminated at no cost and with output unchanged at q_n^A. In this extreme case the MAC^A curve in quadrant (c) would coincide with the horizontal axis. This extreme situation shows the idealized impact of technological change, but it is clearly unrealistic. If processes could be switched costlessly why should polluters pollute in the first place? As we said at the beginning of chapter 2, the argument that pollution is entirely the result of perverse polluter behaviour does not stand up to scrutiny in the real world: much pollution (but not all perhaps) does save the polluter *some* resources.

Let us return to firm A's resource-using switch of processes as it abates from E_n^A to E_f^A. As the process curve in (b) moves from $P1$ to $P2$ the level of marginal social cost at each output falls from MSC_1^A (q_j^B) to MSC_2^A (q_j^B) in (a). The fall in the marginal social cost at any output is the reduction in marginal pollution cost, at that output, attributable to the new technology, minus the marginal cost of the process switch. But there are also social gains to be obtained from abating further, from E_f^A to E_j^A, this time by reducing output from q_n^A to the level q_j^A, where we have $P^A = MSC_2^A$ (q_j^B) in quadrant (a) and $MAC^A = MPC^B$ (q_j^B) in (c). The implications of a flexible technology can be seen by comparing this solution E_j^A, q_j^A, with the fixed technology solution E_2^A, q_2^A, which is shown in figure 3.1 and inserted in figure 3.4. The availability of process-switching as a lower-cost means of abating than reducing output (over *some* range of abatement) leads to a *higher* socially efficient level of the polluter's output, and a *lower* socially efficient level of pollution. The more flexible is technology, that is the cheaper is a process switch, the weaker is the trade-off between pollution control and output/employment which we observed in the case of a rigid technology (p. 63). In evaluating the likely impact of pollution control on a polluter, therefore, it is essential that the pollution control agency has information on the availability and cost of less polluting processes. Only with such information to hand can the agency counter the standard objection to pollution control that it will lead to a large reduction in output (and employment) in the polluting industry.

This examination of efficiency under a flexible technology has assumed that the polluter has a number of alternative production processes available for use at a particular location. Another situation that may arise is the availability of a number of alternative locations in which a given process may be employed.[31] The likelihood of a number of locations being competitive in cost terms is greatest when polluting activities are at the planning stage. Hence the interest in the possibility of using zoning regulations to control the location of new polluting activities (see section 4.2.3 below). The choice of location is important from the pollution point of view only if alternative locations offer differing levels of pollution cost for any level of the polluting activity. Such differences could arise from regional differences in the distribution of pollutees and of other activities whose pollutants may interact with the one in question. In other words, a switch of location *may* offer another means of breaking the relationship between output and the amount of pollution cost incurred which is analogous to a switch of processes. In contrast to a process switch a change of location does not reduce the quantity emitted to the environment generally; rather, it reduces the quantity emitted to that part of the environment that is utilized by pollutees. Typically, for example, the relocation of a water-polluting firm further downstream does not reduce the quantity of effluent discharged to a river, but it does reduce the amount emitted to the stretch of river used by others, and thereby reduces the pollutant concentration of that stretch. Viewed in this way, relocation reduces the level of E^A, where E^A is the quantity of pollution affecting the pollutees. Consequently the polluting firm A, faced with a choice from numerous locations, each offering a different level of E^A for any level of q^A, will place them in order from most to least pollution-intensive. To keep to the conventional model let us assume that the marginal cost of moving to a particular location is greater the less polluting is the new location; in other words, the marginal abatement (moving) cost curve slopes downwards.[32] The choice of location can then be analysed along the lines of the choice of process when technology is flexible. In figure 2.3 (p. 20) the $P1$, $P2$ etc. rays could be different locations (where the E variable is interpreted as the quantity of pollution affecting the pollutees); and in figure 3.4 the abatement from E_n^A to E_j^A derives from a move from the initial location to a location corresponding to the pollution-output curve $P2$ in quadrant (b). This abatement is followed by abatement from E_j^A to E_j^A through a reduction in output from q_n^A to q_j^A in the new location.

The introduction of the problem of a choice of locations does not alter the basic reasoning of the efficiency analysis. But it does mean that, ideally, the search for socially efficient solutions should cast the net wide to include a comparison of *all* available means of abatement: cutting output, altering technology and changing location.[33] Once the least cost abatement method (or combination of methods) has been identified, the socially efficient solution is the one that presses this least cost method to the point at which the extra cost of further abatement outweighs the extra benefit of the pollution cost eliminated.

(4) *Polluting monopolists*

The literature concerning the effect of the external costs of production on allocative efficiency has mainly used a competitive market analysis, both polluters and polutees (where these are firms) being assumed to operate in perfectly competitive goods and factor markets.[34] The important question that arises is whether it is likely that this analysis of efficiency, in particular an analysis of the efficiency of pollution control *policy*, is likely to yield misleading conclusions for those cases in which pollution derives from imperfectly competitive markets. There is some difference of opinion on this matter. Professor Buchanan has argued that the competitive analysis *is* likely to be misleading because a reduction in pollution, resulting from the internalization of external costs, which is socially efficient under competition, may be socially *in*efficient if the polluter is a monopolist.[35] Professor Baumol, on the other hand, has asserted that the criticism of the competitive analysis is not serious because the competitive model is appropriate to the pollution involving large numbers which is the main source of concern.[36]

Let us consider these two views, beginning with Baumol's because it can be dealt with rather briefly. The question at issue is whether the analysis is affected by the polluter being a monopolist in his goods market. The fact that a particular type of pollution may involve a large number of polluters (and probably pollutees) is really not germane to this question. It is quite possible that for a given pollutant there are a large number of firms from different industries who are polluters. Each of the firms may have some market power in its goods market but still be one of many sources of the pollutant. In a large city there may, for example, be many sources of air-borne sulphur oxides and particulates, but one would hardly take this as evidence of the competitiveness of the goods markets that the firms operate in, unless of course the city only contained a few industries.

Buchanan's argument cannot be dismissed so easily, so we must outline its main points and see under what conditions the implications could be significant.[37] His analysis assumed that the polluting firm, A, is a monopolist whose technology and location are rigid, so that a cut in output is the only available means of abatement. Given these assumptions it can be shown that if the monopolist does take account of the whole social cost of his pollution, and abates through cutting output, then social losses may result. In other words, the outcome may be socially inefficient. As we shall see, the basic reason for this is that the monopolist tends anyway to produce an output that is below the socially efficient level, and abatement may involve a move further away from this efficient output level.

If a monopolized industry pollutes unhindered it will choose the output at which marginal private cost equals marginal revenue, $MFC^A = MR^A$, which is $Q_n^A(m)$ at point f in quadrant (a) of figure 3.5. Producing this output with the production process represented by the $P1$ line in quadrant (b) generates the pollution level $E_n^A(m)$ in quadrant (c). These output and pollution levels are lower than those that would occur if the industry were competitive, $Q_n^A(c)$ and $E_n^A(c)$ in figure 3.5, because competitive firms press output to the point at which $P^A = MFC^A$.[38] If the monopolist with a fixed technology were induced to take account of the external cost of his pollution, thereby facing the marginal social cost curve $MSC^A(q_2^B)$ in quadrant (a), he would abate from $E_n^A(m)$ to $E_2^A(m)$ in (c) by reducing output from $Q_n^A(m)$ to $Q_2^A(m)$ in (a), where $MSC^A(q_2^B) = MR$.[39] Buchanan's point is that the move from $Q_n^A(m)$ to $Q_2^A(m)$ may involve a net social *loss*. The social benefit of the output lost (that is $Q_n^A(m) - Q_2^A(m)$) is the consumers' evaluation of the

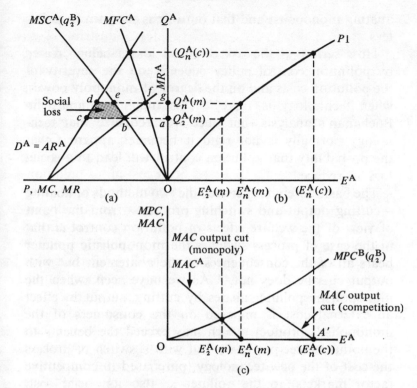

FIGURE 3.5 *Efficiency and the polluting monopolist*

output, represented by the area under the demand curve *acdg* in quadrant (a). The social cost of the output $Q_n^A (m) - Q_2^A (m)$ is the area under the social cost curve *abeg*. As the curves have been drawn the cut in output from $Q_n^A (m)$ to $Q_2^A (m)$ leads to a net social loss equal to $bcde = acdg - abeg$. Therefore controlling a monopolist's pollution *may* not be a socially efficient thing to do! With the monopolist's output being below the socially efficient level already,[40] the further reduction of highly valued output causes a loss to be incurred by consumers that is ignored by the abating monopolist and that outweighs the gain to pollutees.

How serious is the risk of social losses being created by pollution control policy depends on the severity of the pollution costs and on the degree of monopoly power, when technology is rigid.[41] This much is implicit in Buchanan's analysis, but once we recognize that technology normally is not rigid it becomes apparent that the probability that pollution control will lead to a social loss is reduced.

The basic difference between the two methods of abating – cutting output and switching process – from the point of view of the welfare effects of pollution control is that in the case of process switches the monopolistic polluter bears all of the consequences of the abatement but with output cuts he does not.[42] As we have seen, when the monopolistic polluter abates by cutting output in effect an external cost is imposed on the consumers of the monopolist's product, which may exceed the benefits to the pollutees (see pp. 76–7). But with a switch of process the cost of the new technology (purchased in competitive factor markets) to the polluter is also its social cost. There are no external costs to the abatement by switches of process. We may presume, therefore, that abatement through process switching always yields a net (social) benefit, otherwise the monopolist, who pays the cost and reaps the benefit, would not undertake it.

It is evident that Buchanan's objection to pollution control policies relies on the external cost of ouput cuts. Consequently if the technology is flexible, that is, process switches can be made at low cost, his objection can be viewed more as a theoretical curiosity than as a realistic objection to pollution control. To illustrate the point in relation to figure 3.5, consider the extreme case of costless process switches. The monopolist who wishes to abate from the initial pollution level E_n^A (m) adopts processes represented by output-pollution lines lying to the left of $P1$ in quadrant (b). This abatement pivots the marginal social cost curve MSC^A (q^A_B) towards the marginal private cost curve MFC^A. In the limit the costless switching leads to an ouput-pollution line along the Q^A axis in quadrant (b) with the pollution level falling to zero, and output remaining at Q^A_n (m) in quadrant (a). Clearly there is no source of social loss, and society benefits to the extent of the pollution cost eliminated, which is the area Oef in quadrant (a).

The situation is naturally less clear-cut when account is taken of the costs of process switches. It is possible for the net social loss from an output cut to outweigh the net social gain from a process switch when abatement is achieved partly by each method. But the closer pollution abatement comes to being achieved entirely through a change in technology, the greater is the likelihood that overall the abatement will yield net social benefits.

Even in those pollution cases where the degree of market imperfection and the rigidity of technology give rise to real fears of a net social loss resulting from the internalisation of external pollution costs, it is not necessary to adopt Buchanan's nihilistic stance concerning the benefits of pollution control policy. His argument that the *possibility* of net social losses is damaging to the case for cost-internalising taxes (on such taxes see section 4.2.1 below) and supports the case for freely operating markets ignores two considerations. First, if the technology is fixed even free market solutions to pollution problems may lead to net social losses, because they also operate as a method of internalising external costs and would induce output cutting by the abating polluter (see page 95 below). So it is not only the interventionist who must confront the possibility of perverse effects of pollution abatement. Second, a *more* interventionist policy such as issuing directives to polluters on the pollution abatement technology to be adopted, or offering subsidies for such technology, may be able to reduce the risk of social loss by biasing the selection of abatement method away from the socially risky reduction in output.

The conclusion of this discussion of polluting monopolists is that the possibility that pollution control may lead to a social loss cannot be ruled out on theoretical grounds. A loss is least likely to occur, however, where the monopoly power is weak, the pollution cost is not severe or the technology is flexible. In particular a pollution control agency could afford to ignore the problem raised by Buchanan in cases where empirical studies of polluting industries (or consumption activities) showed that a change in the *form* of the activity would be cheap relative to a curtailment of the *amount* of the activity.

3.2   PROPERTY RIGHTS AND MARKET SOLUTIONS

In the last twenty years economists have made a substantial, and arguably excessive, effort to identify the characteristics of theoretical market solutions to pollution problems. The main result of this effort has been a lengthy list of the likely obstacles to the achievement of socially efficient outcomes without government intervention. The problems facing the interventionist will be considered in chapter 4; in the remainder of this chapter we first briefly describe how an idealized market would work, and then summarize the list of obstacles to the operation of markets in the real world.

3.2.1   *Idealized markets*

In the context of the solutions to problems of external cost, the term 'market' generally refers to private bargains between polluters and pollutees. It is assumed that bargaining takes place when the property rights of the parties involved have been determined by the legal system or by the government. It is important to be clear at the outset that the property rights invested in one party or the other are of a specific kind: they are the right to the enjoyment of the *net benefit* obtainable from a production or consumption activity in the absence of interference (damage) by another activity. The relevant net benefit is the *profit* from production or the *net utility* (consumer's surplus) from consumption.

If a bargain between the parties is to involve a departure from a position of freedom from interference for the party who possesses the property rights, then compensation for damages will have to be paid by the other party, which is at least equal to the loss of net benefit sustained.[43] We shall see in a moment, however, that efficiency requires the payment of this equivalent compensation in a particular form when the pollutee possesses the property rights (i.e. when the polluter is liable). The establishment of polluter liability effectively provides the pollutee with freedom from uncompensated loss. No-polluter-liability permits the polluter to impose such uncompensated damage.

Let us see how a bargain might work once property rights are established, and then examine whether the assignment of rights will affect the nature of the bargaining solution. Consider first the case in which the polluter, A, has a fixed technology and the pollutee, B, possesses the property rights. In the absence of an agreement with B over compensation for pollution cost incurred, A would have to avoid imposing uncompensated harm by stopping producing and polluting. Consequently the zero points in figures 3.1(a) and 3.2(a) can be thought of as the starting point for negotiation with polluter liability. If bargaining were costless a mutually advantageous move away from this point could be negotiated. For a small move away from zero pollution in the two figures the initial benefit to A is shown by the high level of marginal abatement cost, MAC^A, the amount of profit obtainable from A's first unit of output. The initial cost to B from the first unit of output by A is shown by the low marginal pollution cost, MPC^B, level. It is apparent that *if firm B were producing the output* q_2^B, and were therefore facing the marginal pollution cost curve MPC^B (q_2^B) in figure 3.2(a), the first unit of A's output yields a benefit to A in excess of the cost to B. If A offered B just enough compensation to cover the pollution cost, then B would not lose and A would gain from an agreement to permit this first unit of output. The net gain from the bargain, the distance between the MAC^A and MPC^B curves in figure 3.2(a) at the first unit of output, would be shared between A and B. Given that B possesses the property rights, perhaps B will try to capture a large share by posing as the reluctant negotiator, but however the gains are shared there is a mutual benefit to agreeing to permit both the first unit of output and the subsequent units up to q_2^A.

At output q_2^A the maximum offer by the polluter, equal to MAC^A, just equals the pollutee's marginal pollution cost. By just making that offer the polluter is implicitly taking into account, in his output choice, the full marginal social cost of his activity. Above q_2^A there are no further mutual gains, because the extra compensation required by B to cover the marginal pollution cost exceeds the marginal benefit to firm A from the further expansion of its output. The process of costless bargaining appears to lead to the socially efficient level of q^A under polluter liability. But the critical reader will have noticed a crucial, but unsubstantiated, assumption of the above description of the bargaining process: that the pollutee, firm B, negotiates as if its own output were fixed at the socially efficient level q_2^B. Recognizing the unlikely nature of this assumption brings an interesting problem to light.

A more appropriate designation of the starting point for negotiation is q_0^A, q_0^B in figure 3.1(a) and (b), rather than the q_0^A, q_2^B previously assumed. As it is producing at the *higher* level q_0^B at the start of negotiation, the pollutee firm B regards the MPC^B (q_0^B) curve in figure 3.6 as the initially relevant pollution cost curve in deciding the amount of compensation. Suppose that firm A were to raise its output from zero towards the socially efficient

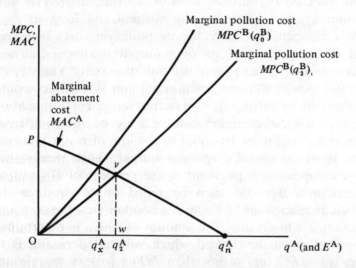

FIGURE 3.6 *Inefficiency with open-ended polluter liability*

q_2^A in quadrant (a) of figure 3.1. We know that, in the absence of compensation, B would face a marginal cost curve that pivoted from MFC^B (q_0^A) eventually to MFC^B (q_2^A). This increase in cost would induce the pollutee to adjust his output towards the socially efficient level q_2^B, as figure 3.1, quadrant (b), shows. Simultaneously the polluter A would adjust towards the socially efficient level of his output as it operated on the marginal pollution cost curve MPC^B (q_2^B) in figure 3.6. But imagine the pollutee's response if, as a consequence of the bargaining over A's pollution, it receives equivalent compensation from the polluter. The pollutee now has no incentive to cease operating on its original marginal private cost curve MFC^B (q_0^A), since the pollution cost is balanced by the compensation. The pollutee behaves as if there were no pollution and retains its high output level q_0^B. In the bargaining process B treats the curve MPC^B (q_0^B) in figure 3.6 as its minimum compensation curve, and the bargain will not proceed beyond the level of A's output q_x^A. The outcome q_x^A, q_0^B is not socially efficient. As long as firm B is compensated fully for its pollution costs incurred, at whatever level of its own output it chooses, it will have no incentive to minimize pollution costs. Yet a move to the socially efficient position requires B to reduce pollution costs by cutting *its own* output to q_2^B. The implication is that, if social efficiency is to be achieved, the compensation payment must be equal to the loss of profit incurred by B on its *socially efficient* output. This means that the compensation payment is independent of B's output level, and B could then be relied on to produce the socially efficient q_2^B.[44] From B's point of view the compensation is a lump sum, the amount of which is determined by A's output level, and which will not persuade B to regard MFC^B (q_0^A) rather than MFC^B (q_2^A) as its relevant marginal cost curve. The bargaining process would then be as we initially described, leading to q_2^A, q_2^B. In the case of polluter liability, therefore, social efficiency will be attained through costless bargaining only if property rights are defined so as to limit the polluter liability to the pollution cost associated with the socially efficient level of B's output; open-ended liability can lead to inefficient outcomes.

When the polluter has the property rights, and is not therefore liable for compensation, the starting point for negotiation is q_n^A, q_n^B in figure 3.1, quadrants (a) and (b) and q_n^A in figure 3.2. The polluter produces and pollutes up to the point at which there is no extra benefit from further output, unless an agreement is reached with B to curtail output in exchange for a compensation payment to the polluter by the pollutee. Curtailing output imposes a cost on A: it is the opportunity cost of the foregone output, in other words the loss of profit. The MAC^A curve shows, for each level of q^A and E^A the minimum amount of compensation that firm A will accept if it is to agree to reduce its output. In order to give up the q_3^A unit of output, for example, A would require (extra) compensation at least as high as yz in figure 3.2 (a). In considering a bargain from the q_n^A starting point the pollutee will regard it as worthwhile offering to pay to A, at any level of q^A, compensation up to the level of marginal pollution cost. At output q_3^A it will pay B to offer compensation up to xz in exchange for the elimination of the q_3^A output unit. Starting from q_n^A, abatement through a reduction of A's output will yield mutual benefits until output q_2^A is reached. Once again the share of each party in the net benefit from the bargain will depend on bargaining strength; and in this case the polluter, who is not liable, may hold out for the lion's share. The series of offers by the pollutee induces the polluter to take account of the *social* cost of his activity. Abatement below q_2^A will not yield any extra benefit to be shared, so that with the assignment of rights to the polluter the costless bargaining process leads to the socially efficient level of A's output, q_2^A. With this assignment of rights the compensation payment, made by the pollutee, need not be in a lump sum form. The bargain provides A with the incentive to move to q_2^A and this move will automatically induce B, who receives no compensation, to select the socially efficient output q_2^B.[45]

The essence of this analysis of bargaining solutions is that the allocation of resources, between A's and B's activities, is independent of the assignment of property rights if the negotiated payments are in the right form. With either assignment of rights the socially efficient allocation, producing the outputs q_2^A and q_2^B, will be achieved by a market in its idealized form. This allocative-symmetry proposition is known as the *Coase theorem*.[46] It should be noted, in view of our emphasis on the realism of a flexible technology assumption, that the analysis of idealized bargaining solutions does not require the polluter's technology to be fixed. If we substituted a flexible technology marginal abatement cost curve in figure 3.2, the bargaining procedure would be much as before except that the *form* of abatement that is the result of the negotiated level of pollution would include some switching of processes. Naturally the polluter's socially efficient output level would be higher than under a fixed technology; but the ability of bargaining to bring about the socially efficient output and pollution levels from either assignment of property rights would not be affected.

Although the allocative effects of an idealized market are symmetrical with respect to the assignment of property rights, the outcome is not symmetrical in other respects. In particular, the assignment of rights determines the direction of the flow of compensation. The wealth of the polluter or pollutee is at least protected, and more probably raised, by possession of the rights. If the polluter is liable, then the negotiated pollution level q_2^A in figure 3.2 (a) leaves the pollutee B no worse off than if there were no pollution and probably better off, since B will take a share of the net social benefit Opv (i.e. B will be over-compensated). The polluter will also be better off than if there were no pollution to the extent of his share in the net social benefit, but will not receive the Ovw part of the profit from q_2^A because this (at least) it must pay out in compensation. If the polluter is *not* liable then, compared with the zero pollution situation, it is better off at q_2^A to the extent of the full profit from that output, $Opvw$, plus the compensation received from the pollutee to refrain from producing above q_2^A (which will need to be at least wvq and will probably be greater since it will take a share in the net social benefit derived from the abatement from q_n^A to q_2^A[47]). Clearly, the polluter will be better off if it is not liable than if it is. With no polluter liability the pollutee, compared with the no-pollution state, is worse off at q_2^A to the extent of the pollution cost incurred, Ovw, and the compensation paid, at least wvq. Certainly the pollutee's wealth (or its welfare if there is no negotiation and it simply bears the pollution cost) is lower if it fails to obtain property rights than if it succeeds.[48] From this discussion it is clear that the assignment of property rights does affect the distribution of wealth. There are two aspects of this asymmetry that could be important in practice in deciding who should have the rights. First, if pollution tends to affect lower-income people proportionately more than higher-income people, either directly or through the impact on goods prices and profits, then opting for no polluter liability will have a regressive effect on the distribution of welfare.[49] Second, quite apart from the distributive impact of pollution, if the imposition of uncompensated harm is regarded as unjust, then polluter liability is required to provide just protection for pollutees. One of the implications of the Coase theorem is that, because *either* assignment of property rights can lead to social efficiency, we may feel free to make the assignment on other criteria, including the desire for just protection. Unfortunately, however, life is not that simple; as we shall see in chapter 4 we cannot assume that in the absence of idealized markets, efficiency and justice are mutually compatible; and the design of a pollution control policy that is both efficient and just in some degree is a complex problem. But first let us review the obstacles to perfectly operating markets in external costs.

3.2.2 *Obstacles to Socially Efficient Market Solutions*

If for some reason private bargains over pollution levels are unable to generate socially efficient outcomes, we shall refer to this as 'market failure'.[50] A special case of market failure is the inability of markets to internalize external costs at all. Ths list of obstacles to the 'success' of markets will be classified into those that may arise even if the number of polluters and pollutees involved is small (many of which apply *a fortiori* when there are large numbers involved) and those that are associated only with large numbers.[51]

In general the obstacles that will be outlined are relevant whichever assignment of property rights is chosen. In one or two cases it can be argued that the obstacle is more serious under one assignment than the other, as we shall note in due course, but the implied challenge to Coase's allocative-symmetry proposition will not be investigated in detail. If there are, for many pollutants, insurmountable obstacles to socially efficient bargains *whoever* possesses the property rights, the allocative-symmetry issue reduces to one specific question: if *no* bargain can be struck, which assignment of rights leads to the less socially inefficient outcome?[52] The answer requires a comparison of the alternative 'starting' points in figure 3.2 (assuming for simplicity that B produces $q_{\frac{B}{2}}$). Compared with the socially efficient state, $q_{\frac{A}{2}}$, the finishing point $q_{\hat{0}}^A$ foregoes the social benefit Opv, whereas the finishing point q_n^A foregoes vnq. Which is the larger loss depends on the slopes and positions of the two curves. The more seriously damaging are increases in the emission of the pollutant the steeper will be MPC^B ($q_{\frac{B}{2}}$) and the greater will be the loss at q_n^A (no polluter liability) relative to the loss at q_0^A (polluter liability). Certainly there is no reason to expect the social efficiency of the 'no-bargain' outcome to be the same under both assignments of property rights. Consequently the placement of liability is likely to matter – from the point of view of both social efficiency and justice.

(1)   *Obstacles even with small numbers*

Non-marginal bargains Unless the bargaining process takes a particularly demanding form there is no guarantee that a socially efficient solution will be agreed upon. The problem is that from either starting point q_0^A or q_n^A in figure 3.2, there are numerous mutually advantageous bargains to be struck. If, for example, the polluter is liable, what force is there to drive the bargain to q_2^A rather than to some other output level such as q_1^A? If bargaining were concerned with each successive *marginal* unit of A's output it would be apparent at each step, until q_2^A is reached, that a joint benefit can be obtained. But it is hardly plausible to assume that the smoky steel factory owner will ask the laundry firm how much it will accept for the first cubic metre of smoke to be allowed, how much for the second, and so on. The bargain is more likely to be over how much the pollutee requires to be compensated to leave the polluter free to pollute (at output q_n^A). The answer is that it will want at least *Onq* in figure 3.2 (a), and the polluter may decide either to pay and pollute (if *Opq* > *Onq*) or not to pay and to stop polluting altogether (if *Opq* < *Onq*). This bargain over the *total* benefits of polluting and not polluting will not lead to the socially efficient output q_2^A.

Threat-making The idealized bargaining process, which envisages the adjustment to socially efficient outputs, requires that neither the polluter nor the pollutee has an incentive to use threats to increase its return to a bargain. In practice the polluter firm that is not liable might exaggerate the benefits that it derives from its output, and threaten to produce above q_n^A. Not knowing the polluter's true MAC curve, the pollutee would fear the imposition of extra pollution costs, and offer compensation for the abatement of the *threatened* pollution from the output above q_n^A. The bargaining would take place on the basis of the polluter's fictitious marginal abatement cost curve, which would lie above MAC^A in figure 3.2 (a), and even if bargaining were to proceed over marginal units of abatement the agreed output would exceed q_2^A. A number of highly implausible reasons for believing that threat-making will not prevent the attainment of social efficiency have been offered by the devotees of idealized markets.[53] Rather than pursue this red herring, let us recognize that threat-making is partly the result of imperfect information. Clearly, the threat of extra polluting output will be less convincing if the pollutee knows that this extra output would reduce the polluter's profit level. Even so, it will not be entirely unconvincing unless the pollutee is sure that the polluter is a pure profit-maximizer devoid of malicious intent.

Poor information   Even without threat-making as a likely mode of behaviour, bargaining in its idealized form requires the pollutees who are potential negotiators to be fully aware of the consequences of the pollutant, including its impact on health, amenity and production. With residuals that are insidious this information may be costly to obtain (particularly on a decentralized basis by individual pollutees rather than by a government agency) or even completely unobtainable given the current state of scientific knowledge. It is hard to imagine that bargaining could take place over any but the most obviously damaging pollutants, whoever possesses the property rights. Not only do the pollutees need to know their pollution cost curve; if they are to negotiate they must be able to trace the *sources* of the damaging pollutants. Even when the number of polluters is relatively small this can be difficult, as may well be the case, for example, with the secret dumping of poisonous chemical wastes.[54] When there are multiple and often distant sources, the identification of the relevant polluters will be beyond the capabilities of most pollutees. There is then little likelihood of bargains leading to even approximately socially efficient solutions in these circumstances.

Transactions costs In addition to the cost of obtaining information, any bargaining process involves the use of resources such as the time of the negotiators and the labour of legal and technical advisers. Depending on the simplicity or complexity of the case these may be small relative to the potential benefit from a bargain, as perhaps in the case of an agreement between neighbours to keep noise to a minimum at certain times, or they may be large, as in cases where there is a technical dispute over the sources and consequences of pollutants or a litigious dispute over liability. In general one would expect the costs to increase (perhaps exponentially) with the sizes of the polluter and pollutee groups; it has been argued also that transactions costs will be higher if the polluter group is not liable than if it is.[55] If this were so then it is possible that transactions costs could provide an obstacle under the no-polluter-liability regime even when they wouldn't be a problem with the polluter liable. The question is, why should the assignment of property rights affect the costs of reaching a settlement, in or out of court? If the polluter is not liable the onus is on the pollutees to identify other members of the affected group, to reach an agreement on a unified policy if they are to act as a group, and to identify, contact and bargain with the polluter. If the polluter firm is liable, on the other hand, it must initiate the bargaining (in order to avoid a prohibition order or injunction, on his polluting activity) by identifying, contacting and bargaining with the pollutee group; and the pollutees will still incur their collaboration costs even though they are responding to, rather than making, a bargaining initiative. It is not clear that there is a significant asymmetry with respect to bargaining costs. whether these costs are generally high or low.[56] Consequently, in those cases where the obstacles to bargaining are not insurmountable, probably small numbers cases where the nature of pollution costs is obvious, the assignment of liability could simply be made on grounds of justice.

One point should be made clear in connection with information and transactions costs. If such bargaining costs are high, for example in excess of *Opv* under a regime of polluter liability in figure 3.2(a), they eliminate the social benefits from a bargain which could theoretically have been achieved through costless bargaining. Since no social benefits are then available from bargains we cannot say that bargaining fails to achieve the socially efficient output combination q_2^A, q_2^B. This output combination is no longer social efficient *as far as bargains are concerned*. However, if there are other means of altering the allocation of resources that *can* reap social benefits from a modification of the polluter's activity, such as some form of pollution control policy, then we can say that there is a *relative* market failure.

The problem of negotiated compensation payments We saw in section 3.2.1 that the pollutees will select the socially efficient level of their activity only if any compensation payment received by them relates only to the pollution costs they incur at that level of activity. As we have seen, this requires the polluter to be subject to

limited liability, which effectively makes compensation
a lump sum from the pollutee's point of view.[57] If property
rights were defined in this way costlessly (and the various
other obstacles to bargains were absent) the bargain could
proceed to generate social efficiency. However, the rather
stringent requirement that the socially efficient level of
the pollutee's activity be identified *before* bargaining takes
place means that in many cases costly legal involvement
will be necessary. Imagine for example attempting to
define the limits of polluter liability for an air pollutant
in a city with a multitude of different affected activities.
Disputes over the *degree* of liability weaken the general
case in favour of markets, that they permit decentralized
decisions without the heavy involvement of centralized
institutions. The point is that when liability is disputed,
and this is more likely if the liability is limited than
if it is open-ended because the polluter has a possible
legal means of reducing the extent of his commitment,
judicial involvement is likely to become a major component of the cost of controlling the pollutant. The idea
of a market operating with a general background definition of the rules of the game offered at low cost by
the courts becomes implausible when rights are hard to
define, especially when there are many potential litigants.

Possible myopia of bargaining   We have seen that as a
rule bargaining needs to be concerned with the costs
and benefits of *marginal* units of pollution if social
efficiency is to be achieved. This simple view becomes
somewhat clouded when we admit the possibility that
the marginal pollution cost curve slopes downwards.[58]
For in this case a costless bargaining process proceeding
incrementally may lead to a socially inefficient position.
Consider, for example, a regime of polluter liability in
figure 3.3, quadrants (b) and (e). In quadrant (b) the
starting point for negotiation is q_0^A, and when bargain-

ing takes place over the first unit of q^A it becomes apparent that the compensation required by the pollutee, B, exceeds A's maximum offer since $MPC^B > MAC^A$. No bargain results, and if bargaining is sequential, with the second unit of q^A being considered only after the first unit is agreed upon, with appropriate compensation, the points q_k^A in (b) and q_r^A in (e) at which social benefits begin to accrue will never be reached and the polluter must cease polluting. In quadrant (b) this means that the socially efficient level of A's output, q_i^A, is not achieved. In quadrant (e), on the other hand, the same result happens to be socially efficient.

If, instead, the polluter is not liable, and q_n^A is the starting point, the pollutee will make an offer for the first unit of abatement that exceeds the polluter's abatement cost, and in figure 3.3 a move towards q_i^A in (b), or q_s^A in (e), will occur. But q_i^A is socially efficient whereas q_s^A is not, so that once again marginal bargaining will not be certain to lead to social efficiency. The basic problem is that marginal bargaining is myopic: it is as capable of leading the system to a *local* maximum such as q_s^A in (e) as to a *global* maximum (socially efficient point) such as q_i^A in (b). On the other hand, bargaining over the total benefits and costs of polluting has its own problems.[59] Ideally, bargaining should be marginal if there are social benefits to be obtained on a particular unit of the polluter's output, but it should take the broader view where there are not, so that benefits elsewhere can be spotted. It is hard to see, however, how a bargaining procedure can hope to combine these ideal qualities. It is not sufficient for the advocates of bargaining simply to assert that any available social benefits will be recouped by bargains;[60] the doubters will take more convincing than that!

Bargains with polluting monopolists The idealized bargaining process described earlier necessarily brings about a socially efficient allocation of resources if the goods and factor markets are competitive, but not if there are market imperfections. This follows from the earlier analysis of a polluting monopolist:[61] the purpose of bargaining is to induce the polluter to take account of the *social* cost of his pollution. Yet if the monopolist is thereby encouraged to abate by cutting output, a social loss may result, as we have seen. This is most easily shown for the fixed technology case in figure 3.5. The analysis of idealized bargaining has shown that the existence of mutual benefits to a bargain would lead to the pollution level $E_{\frac{1}{2}}^A(m)$ in quadrant (c) at which $MAC^A = MPC^B$. In order to reach this pollution level the monopolist selects the output level $Q_{\frac{1}{2}}^A(m)$ which, compared with the unhindered profit-maximizing output $Q_n^A(m)$, is associated with the social loss *bcde* in quadrant (a). This problem with bargaining solutions would clearly not apply in cases where controlling the monopolist's pollution leads to net social benefits.[62]

(2) *An extra obstacle peculiar to the large-numbers case*

Many of the obstacles that have been mentioned so far become even more serious when there are large numbers of polluters and/or pollutees, but there is one obstacle that becomes relevant only when large numbers are involved. This is the problem of strategic behaviour and free-riding. Consider for example a smoke nuisance imposed by one factory on a large group of people. If the polluter is liable, and the pollutees wish to bargain as a group to strengthen their position for compensation negotiations, then the pollutees must decide on the group's minimum and maximum compensation levels for different levels of pollution. Ideally these compensation levels should be the sum of the minimum and maximum levels for all members of the group. But when the members are asked to reveal their personal compensation requirements they will be tempted to exaggerate because they will expect that the eventual share-out will depend on these requirements. Each member will also calculate that the probability of the group reaching an agreement with the polluter will not be reduced significantly by his exaggeration. The result of this strategic behaviour is that the bargain, if any is reached, will be based on an over-estimate of the pollution costs, and will therefore be unlikely to lead to the socially efficient pollution level.

An analogous problem arises if the compensation payments are made by a group (of polluters under polluter liability, or of pollutees if the polluter is not liable). Imagine, for example, an attempt by a group of pollutees to agree on a bribe to be offered to the polluter to induce him to abate if he is not liable. Each member of the group will calculate that his share of the eventual group payment will be based on his stated maximum offer, and that if he understates his true pollution cost and makes a lower offer the probability of an agreement being reached will not be affected. In other words, each member will be tempted to take a cheap, or even free, ride on a group agreement. Once again it is clear that any bargain that is arrived at will not reflect the true social benefits and costs of pollution abatement, and will not, therefore, lead to a socially efficient allocation of resources.

CHAPTER 4

Pollution Control Policies

4.1 OBJECTIVES: EFFICIENCY AND JUSTICE AGAIN

In section 3.1 above it was suggested that the resolution of pollution problems is inherently a resolution of conflicts of interest. Any policy designed to reduce pollution levels will yield benefits to pollutees and impose costs on another group, usually but not necessarily the polluters.[1] At the same time, any policy that fails to totally eliminate emissions leaves the welfare of pollutees lower than it would be in the absence of pollution unless equivalent compensation for the remaining emissions is paid to them. We shall analyse pollution control policies using the premise that, at least above some 'reasonable' level of interference, the just protection of pollutees requires them to be fully compensated. In this way we force into the open the problems of building just compensatory elements into the pollution control policies that aim to move the system towards socially efficient pollution levels.

The analysis of the efficiency of alternative policies will be presented in two stages. In section 4.2 we shall adopt as an initial working assumption that policy-makers have good information on the abatement cost and pollution cost curves for different types of pollutant. This will facilitate the description of the basic features of the policies without the encumbrance of some real-world complications. However, an analysis that is to be relevant to actual policy decisions must recognize the lack of informa-

97

tion and uncertainty about the future with which pollution control agencies are confronted in reality. Consequently section 4.3 considers policies based on imperfect information. To complete the discussion of policy instruments, section 4.4 outlines the implications of pollution control for the distribution of welfare in society.

4.2   POLLUTION CONTROL POLICIES IN AN IDEALIZED SETTING

One hears from many sources a great variety of proposals for *the* solution to pollution problems. Some of these are more analytically elegant than practically possible, and some favour particular interest groups to the detriment of others (for example the suggestion by local environmentalist groups that the proposed third London Airport should be located Elsewhere as distinct from not being built at all). These myriad policy proposals are mostly variants of a relatively small number of basic policy instruments.

(1) *Taxes or charges* The contribution of many economists to the debate over pollution control policies has been the analysis and advocacy of tax or charge schemes which attempt to capture, through centralized intervention, the advantages of using prices to allocate resources between competing uses. The literature is confusing to the uninitiated because the tax schemes and the effluent charge or pricing-of-pollution proposals are essentially the same thing.[2] This is because the tax usually proposed is on the emissions of pollutants rather than on polluters' output or profits. Sometimes the tax schemes involve the use of the revenue to compensate pollutees, and are often misleadingly referred to as tax-subsidy schemes. We shall use the terms 'effluent tax' and

'effluent charge' interchangeably and refer to schemes involving compensation as tax-compensation or charge-compensation policies.

The charge may be uniform, or it may be different for different polluters; and the level of the charge may either be centrally determined by the pollution control agency, or be set through an auction of polluters' rights-to-pollute administered by the control agency.

(2) *Regulation* Regulations are administratively imposed limits on the amount of effluent that polluters are allowed to discharge without incurring a penalty.[3] The limits may be uniform or may vary between polluters.

(3) *Zoning* In the context of pollution control zoning is a planning device for keeping polluters and pollutees geographically separate. It is naturally easier to apply to new pollution sources and new locations for pollutees than to those that are already established.

(4) *Payments or polluter subsidies* An alternative to the inducement of pollution abatement by imposing penalties on polluters is to encourage such abatement through the offer of a government subsidy to cover part of the cost. Such a policy clearly could operate on its own as an incentive scheme, or in conjunction with taxes or regulation to mitigate the financial impact on polluters of a policy of forced abatement.

To begin with (sections 4.2.1–4.2.4) we shall proceed on the assumption that where pollution is created by firms we can usefully analyse the effects of various policies on a static industry, in other words on an industry *with a given number of firms*. This is certainly a reasonable assumption for a short-run analysis. In section 4.2.5 we shall take account of the fact that different policies may lead to different profit levels for the polluting firms and consequently, if there are no serious barriers to the entry of new firms to the industry (which will be taken to be competitive), to different sizes of polluting industries in the long run.

4.2.1   *Charges*

(1)   *Inducing efficient abatement*

Consider to begin with the characteristics of a charge (tax) imposed on polluters at a fixed rate per unit of pollutant emitted. What rate of charge will induce the cost-minimizing polluter to reduce his pollution to the socially efficient level? The answer is a rate equal to the marginal pollution cost at the socially efficient level of pollution.[4] This is illustrated, for the case of a fixed technology, in figure 4.1, which is a slightly modified version of three of the quadrants of figures 3.1 and 3.2 above. We saw, in discussing figures 3.1 and 3.2, that the socially efficient level of pollution is E_2^A which would result from the level q_2^A of the polluter's output. In the absence of a pollution control policy the polluter A would choose the higher output (and pollution) level q_n^A (and E_n^A), which maximizes its profit π^A (no policy) in figure 4.1 quadrant (c), by polluting up to the point u in quadrant (b), at which its marginal abatement cost is zero. If the government imposes a charge of $Ot = fe$ in quadrant (a) per unit of pollution,[5] the charge is just equal to marginal pollution cost wv at q_2^A in quadrant (b). The polluter is now obliged to operate on a marginal private cost curve inclusive of the charge, MFC^A + charge in (a), which is parallel to the original MFC^A curve but higher by the amount Ot. Clearly, the charge reduces the profit on any unit of output by the amount Ot. Since marginal abatement cost is the loss of profit owing to a reduction in output it is not surprising that the

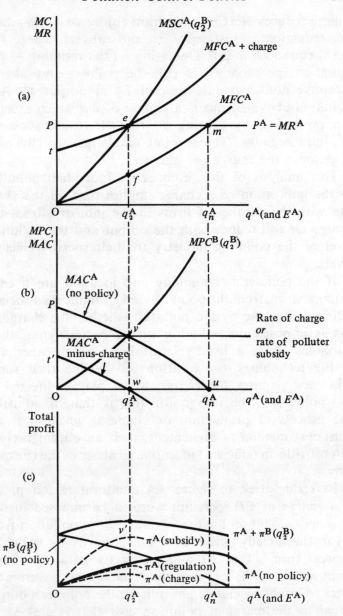

FIGURE 4.1 *A pollution charge (under fixed technology)*

charge reduces MAC^A in quadrant (b) by $wv = t'p$ where the reduction is just equal to the rate of charge ($t'p$ in (b) equals Ot in (a)). Operating on the resulting MAC-minus-charge curve $t'w$ in (b), the polluter now has an incentive not to pollute beyond E_2^A at output q_2^A. Any pollution above E_2^A incurs a charge of wv, which exceeds the profit lost by abating so that MAC-minus-charge becomes negative. The amount of charge actually paid is q_2^A times the charge per unit wv.

This analysis of the response of individual polluters to the imposition of a charge implies that, in the short run with the number of firms in the industry fixed, the charge Ot will reduce both the output and the pollution level of the polluting industry to their socially efficient levels.

If the polluter's technology and location are fixed it is immaterial, from the point of view of inducing a socially efficient response by the polluter, whether the charge or tax is imposed on pollution or on output, since these variables are in a fixed relationship to each other. But if the technology (or location) is flexible then social efficiency requires the charge to be placed directly on the pollution. The reason for this is that, if adjusting the process of production or changing location is the least-cost method of abatement, then an output tax will not provide an (efficient) incentive to abate in the cheapest way.

Referring back to figure 3.4 quadrant (c) on p. 70, if a charge of CD' per unit were to be imposed on the polluter's pollution the firm would reduce pollution from E_n^A to the socially efficient level E_j^A,[6] partly by switching process from $P1$ to $P2$ and partly by reducing output from q_n^A to q_j^A. If, on the other hand, the government were to impose a charge per unit on the polluter's output equal to the marginal pollution cost that is created by the socially efficient output q_j^A when the original technology (P1) is used, the result would not be socially efficient. A charge of TS per unit of output in figure 3.4, quadrant (a), would induce the polluter to reduce output to the level q_j^A, which would be socially efficient under the new technology $P2$, but it would not provide an incentive for the required switch to $P2$. Consequently, in producing the lower output q_j^A while still using the old technology $P1$, the polluter would move part of the way from E_n^A to the socially efficient pollution level E_j^A in quadrants (b) and (c). But its new pollution level, E_t^A, is not socially efficient; at this level marginal pollution cost still exceeds marginal abatement cost.

The moral of this story is that devising socially efficient instruments of pollution control is not simple. A tax on output may be easier to design and administer than a pollution charge, but it does not lead to the socially efficient reduction in pollution unless output cuts are the least-cost method of abatement.[7]

(2) *Compensating pollutees*

A socially efficient level of pollution is determined entirely by the balance of the cost of abating and the damage consequences of not abating pollution. It does not involve any judgement about how much interference pollutees should reasonably be expected to tolerate without compensation. In other words, there is no necessary correspondence between a socially *efficient* level of pollution abatement and a *just* degree of protection of pollutees from uncompensated damage. Let us assume, plausibly, that the socially efficient level of pollution is sufficiently high to cause unjust interference in the absence of any compensation arrangements. The clear implication of this is that the pursuit of social efficiency is compatible with maximum justice only if equivalent compensation is paid to pollutees for any interference that is thought unjust. To keep the analysis simple we shall assume that *any* uncompensated interference is unjust so that a socially efficient solution is completely just only if the pollutees receive full compensation for interference.[8]

If no compensation arrangement is incorporated into pollution control policies, then the pursuit of efficiency necessarily leads to injustice, and the attempt to eliminate pollution in the interests of justice leads to inefficiency. Compromise pollution levels between zero and the socially efficient level might then be chosen in the light of society's preferred trade-off between efficiency and justice. However, it is worthwhile to explore the possibility of devising compensation schemes as a means of releasing the policy-maker from the choice between conflicting objectives. Let us consider the two central questions relating to the provision of compensation to pollutees:

(a)   Does social efficiency require compensation to be paid and, if not, is it compatible with compensation being paid?

(b)   How can the compensation payments be financed?

(a)   The adjustment by the polluter to the socially efficient pollution level is induced by a charge on each unit of pollution equal to the marginal pollution cost at the socially efficient pollution level. It does not require the proceeds of the charge to be paid to pollutees in the form of compensation when pollution is of the 'public bad' type which we have regarded as typical.[9] But would the move to the socially efficient position be obstructed by paying compensation to pollutees from the revenue of the charge scheme? When the charge is set at the level fe in figure 4.1 (a) it is presumed that the pollutees are operating at their socially efficient output, q_2^B. Yet we have seen that if they are compensated fully for the pollution costs that they incur regardless of the level of output they choose to produce, then they will have no incentive to adjust their output to q_2^B.[10] The pollutees are likely to choose a higher output, thereby raising pollution costs and the associated compensation payments. This will reduce the level of pollution chosen by the polluter below the socially efficient E_2^A, if the charge is raised to match the actual marginal pollution cost generated by q_2^A, given that pollutees are over-producing at $q^B > q_2^B$.

However, attempts to compensate those who suffer the consequences of pollution can be criticized on efficiency grounds only if they fail to make the lump-sum payments, i.e. payments that are independent of the pollutees' own behaviour. In practice it will be difficult to distinguish between cases of pollutees who, despite taking care, are suffering high damage levels owing to their high sensitivity, and those who expose themselves to such damage knowing they will qualify for more compensation. The result may be a need to approximate full compensation by defining 'reasonable' sensitivity and 'reasonable' care taken to avoid exposure, and basing uniform payments to pollutees on the damage associated with reasonable pollutee behaviour. This is, of course, exactly what the courts have in mind when they relate damages awards in tort cases to the loss that would have been suffered by the 'reasonable man'.[11]

(b)   Let us assume that the correct charge has been imposed to persuade the polluter to abate to the socially efficient pollution level. Will the proceeds of this charge be sufficient to compensate the pollutees fully (in lump-sum form)? It is easy to show that if, for the representative polluter, the marginal pollution cost rises as pollution increases, the payment of *all* of the revenue from the charge as compensation will over-compensate the pollutee group. The situation for a particular pollutant in the aggregate (from all sources) is shown in figure 4.2. Both the marginal abatement cost and the marginal pollution cost curves are the horizontal summation of the individual

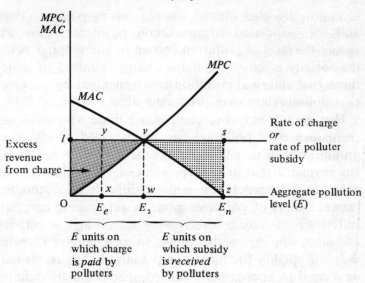

FIGURE 4.2   *Pollution charges and polluter compensation*

polluters' curves.[12] If the rate of charge imposed is wv
the total revenue is $Olvw$. The total pollution cost incurred
by pollutees on the other hand is only Ovw. This is
the simple basis of Dolbear's rather elaborate demon-
stration that the socially efficient charge overcompensates
the pollutees.[13] It is the setting of the charge equal to
the marginal pollution cost at E_2 when the marginal
pollution cost of intramarginal units of pollution is lower
that generates the excess revenue Olv.[14] If the individual
polluter and the aggregate marginal pollution cost curves
were horizontal this would not be so.

It is not obvious, from what has been said, why we
should be concerned about the excess revenue generated
by a pollution charge. From the point of view of justice
a shortfall of revenue would be more problematic, as
we shall see in discussing regulation. And it is possible
to use. surplus revenue to finance the development of

pollution abatement techniques either through subsidies to polluters' research or through research in government agencies. This is not to say, however, that the ability of a charge to generate such revenue, by over-charging on the intramarginal pollution units, is necessarily advantageous. The withdrawal of funds from polluters, which lowers profits below the level that would prevail under regulations yielding similar pollution levels, may have undesirable consequences as we shall see in section 4.2.5.[15] But it is not the impossibility of finding a socially efficient charge rate whose revenue *exactly compensates pollutees* that is the crux of the problem.

4.2.2 *Regulation*

Many economists claim that charges will lead to a more socially efficient solution than will regulation. Before explaining this view we should recognize that in certain respects the two instruments are equivalent in a perfect world of zero cost information, administration and legal enforcement. If a pollution control agency *knows* the socially efficient level of pollution for each polluter then it can equally well set that level as a regulated limit or set a charge that will leave the polluter free to choose that level. If all polluters have similar abatement cost and pollution cost curves, the task is a simple one whichever instrument is used. The only difference in the impact of the two instruments in this idealized world derives from the fact that a charge imposes a financial burden on the polluters for pollution units below the socially efficient level while regulation does not. In figure 4.2 for example a regulated limit of E_2 forces the polluter to bear the abatement cost *wvz*, but the charge *wv* imposes the burden of *wvz and* the tax payment *Olvw*.

If polluters are faced with *different* abatement costs and the discrepancy is known to the pollution control agency, then the regulated limits can be made to differ between polluters so as to minimize the sum of abatement costs across all polluters. Let us explore the logic that lies behind this statement. Consider the situation in which there are two polluters, firm A, which faces heavy abatement costs, and firm C, whose abatement costs are lower. In abating pollution say to the level $\frac{1}{2}E_2^{A+C}$ in figure 4.3, quadrant (a), polluter A would move along MAC^A

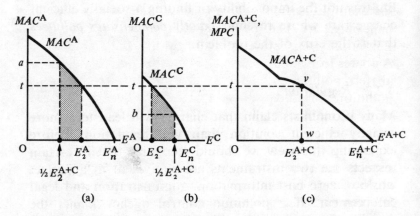

FIGURE 4.3   *Minimizing the sum of abatement costs when these costs differ between polluters*

curve from E_n^A and the last unit of abatement undertaken would incur a cost of Oa. For polluter C to reach the same pollution level $\frac{1}{2}E_2^{A+C}$, starting from E_n^C in (b), the last unit of abatement would impose a cost only of Ob. This is because C's technology is more flexible or its output less profitable. To find the aggregate MAC curve for A and C together, MAC^{A+C}, the individual MAC curves must be summed horizontally. Thus, for polluter A the marginal abatement cost Ot is met at the pollution level E_2^A, and the same cost level is met by C at E_2^C. Consequently, adding together E_2^A and E_2^C, the marginal abatement cost Ot is incurred at the *total* pollution level E_2^{A+C}. If *all* pollution were to be eliminated the total cost of abatement would be the area under MAC^{A+C}, which is the sum of the areas under the individual polluters' curves, MAC^A and MAC^C.

With the aggregate marginal pollution cost curve as shown in quadrant (c), the socially efficient total pollution level is of course E_2^{A+C}. But how should this total pollution level be allocated between the two polluters? Should they, for example, each be allowed to pollute to the level of half of E_2^{A+C}? The answer is that the abatement from E_n^{A+C} to E_2^{A+C} is achieved at the lowest cost if polluter A abates to E_2^A and polluter C abates to E_2^C. The high-cost abater, polluter A, stops abating at a higher level of pollution than the low-cost abater. An efficient allocation of E^{A+C} between the two polluters is achieved if their shares are such as to equate their marginal abatement costs at Ot. In the example shown in figure 4.3, this requires the shares to be unequal ($E_2^A > E_2^C$), as will generally be the case if abatement costs differ between polluters. If, instead, each polluter were restricted to a uniform limit equal to half of E_2^{A+C}, the saving of abatement cost for the low-cost abater (compared with E_2^C) would be outweighed by the extra abatement cost borne by the high-cost abater (compared with E_2^A) as the shaded areas show.

Now if the pollution control agency knows the socially efficient total pollution level, E_2^{A+C}, and knows the abatement costs of the two polluters, then it can simply regulate that A must not pollute above E_2^A and C must not pollute above E_2^C. The outcome of these regulated individual limits is socially efficient. A similar result can be achieved alternatively by setting a pollution charge of Ot per unit; then the polluters will automatically choose pollution levels that satisfy the efficiency requirement that A's share be E_2^A and C's share E_2^C. The point about the charge is that it does not require the pollution control agency to know individual polluters' abatement costs. In the idealized world this is obviously not an advantage. In order to see why it *may* be an advantage in reality we will have to move to a less idealized representation of pollution control policy (in section 4.3).

We should be careful at this point, however, not to conclude that the effects of a pollution charge and regulation are identical in all respects in the idealized world, even though both instruments can generate the same total and individual pollution levels. In the first place, regulation does not have the same revenue-generating capability as a pollution charge, at least if there are no violations of the limits set under regulations. This is due to the absence of a (intramarginal) charge for the pollution units below the limits regulated. Once violations are allowed for, even regulation generates revenue from fines. The magnitude of this revenue depends on the number and size of violations, and on the probability of detection and conviction. All that can be said concerning the size of the fines in relation to the size of pollution costs is that, if socially efficient pollution levels are to be approached, the polluters' *expected* level of fine (taking into account the fact that detection and conviction are not certain) needs to exceed the pollution cost of a violation, otherwise they will have an incentive to violate. Consequently, socially efficient regulation enforcement would generate enough revenue to compensate pollutees for the pollution cost resulting from violations. The compensation of pollutees for the pollution cost caused by the legal, socially efficient pollution is not, by contrast, catered for by fine revenue unless fines significantly exceed the pollution costs owing to violations. Revenue-raising from fines for violations is naturally possible also under a charge scheme, so that we are not talking here about an advantage of regulation over a pollution charge. On the contrary, a charge has the extra source of revenue from intramarginal units of pollution which is lacking under regulation. One way of overcoming the problem of providing for just compensation under regulation is to legislate statutory instruments in such a way as not to discourage common law claims for damages. This requires that the statutes do not pre-empt polluter liability for the pollution costs generated by pollution which does not violate the limits set. Certainly, if regulation is to combine justice with efficiency some attempt must be made to compensate pollutees when the regulated pollution level exceeds the level of pollution that it is just to expect people to tolerate without compensation. The tighter the efficient regulation, of course, the less will be the divergence between efficient and just regulated pollution levels.

Second, while the intramarginal burden on polluters under a pollution charge has advantages from the compensation point of view, it also means (as mentioned before) that the polluter's maximum profit level will be lower than under regulation, as the profit curves in figure 4.1 (c) show. This can have implications for the entry and exit (shutdown) of firms to and from the industry in the long run, as we shall see in section 4.2.5.[16]

4.2.3 *Zoning*

Despite the fact that pollution is usually of the 'public bad' type, leading to the equal exposure of all pollutees in a particular location, there are geographical variations in pollution levels which mean that pollution generally presents a problem of *local* public bads. Many kinds of pollution damage costs are sensitive to the locations of the polluters and pollutees, although the more effective are the residuals-transporting mechanisms of the ecological system, the less true is this general statement. Nevertheless, many of the present-day pollution problems are the result of the concentration of pollution emission combined with the juxtaposition of pollution-generating and pollution-sensitive firms and people.

Zoning is the regulation of the location of polluting and pollutable activities, and it may involve either the enforcement of the *re*location of existing activities, or (more commonly) restrictions on the choice of location for new activities or for activities wishing to move for other reasons.[17] Thus, some areas may be scheduled for industry, others for residential development. Clearly zoning can be regarded as geographical variations in the application of the regulation instrument. In some areas the polluter may be permitted to pollute without restriction; in others he may not be permitted to carry out his polluting activity at all. However, the distinctive feature of zoning from the economic point of view is that it attempts to control total pollution costs through *one* of the determining variables, namely location, while regulated pollution limits that are geographically uniform would operate through another, namely average pollution levels across locations.

The social efficiency of zoning depends on the sensitivity of pollution costs to location choice, on the costs of abatement through relocation or constrained location choice, and (as with other policy instruments) on the costs of administering and enforcing the zoning restrictions. Let us consider the first two of these factors, while retaining the assumption of costless information and administration.[18]

Firstly, for zoning, in its idealized form to offer significant gains to society it is clearly necessary that pollution damage costs be substantially reducable through spatial separation. For certain kinds of pollution this is likely to be the case: noise disturbance, smell and visual disamenity particularly decline rapidly with distance from the source, as do the concentrations of many air pollutants.[19] River pollutants are less directly damaging if pollution-sensitive activities are located upstream of the sources. No doubt if the market could be made to work perfectly by the establishment of property rights in a clean environment, leading to the full compensation of pollutees, then the polluting and pollutable activities would sort themselves into a socially efficient degree of spatial separation.[20] In the absence of such perfect markets but with good information available to the government concerning the geographical variations in the damage resulting from a pollutant, an attempt can be made to identify socially efficient restrictions on location choice. These are not, however, simply those that minimize the sum of pollution damage costs from alternative geographical locations of polluters and pollutees, because these two groups would not be indifferent between locations in the absence of pollution. In other words, restricting location choices is not costless. With given locational differences in damage costs zoning is most likely to be socially efficient if the polluting and pollutable activities are 'footloose' in the sense that they are not location-specific, so that a change of location does not impose high costs. It should immediately be apparent that activities become more location-specific once they are established in an area. It is generally less costly to alter the proposed location of a new activity than to move an existing activity to a new location. For this reason altering location is more likely to be socially efficient at the planning stage than afterwards. Not surprisingly therefore, in practice zoning restrictions have been applied to new activities or new developments of existing activities more frequently than to established ones.[21]

The differential treatment of polluters in different localities therefore offers potential social benefits in certain circumstances. Differential treatment does not, however, have to involve the extreme differentiation of permitting polluters to pollute freely in one area (which has low damage costs) and prohibiting them from polluting at all in another (which has high damage costs). Zoning can be devised so as to establish areas with graded limits on pollution, all of which are socially efficient with respect to their own locality, if the information is good enough. This is true whether the main instrument for pollution control is regulation or a pollution charge: either instrument can potentially reflect regional differences in the target pollution abatement. One advantage of viewing zoning less as a separate instrument to be utilized in isolation than as a complement to regulation or a charge is that the latter aim to control both existing and newly established polluters. With a tightening of controls in *all* zones, but with some zones less tightly controlled than others, the control policy becomes less open to the criticism that it is a means only of controlling new pollution sources. If this approach is combined with a pollutee compensation scheme this would weaken the other criticism that can be made of zoning, namely that it allows uncompensated harm, particularly in areas where controls are less tight (often the districts inhabited by low-income families[22]).

4.2.4. *Polluter subsidies*

The main purpose of regulation, charges and zoning is to induce a movement towards socially efficient pollution levels by imposing on polluters penalties for continuing to pollute to the *privately* efficient levels. The presumption is that it is just to require polluters to bear the full social cost of their activities. While many people, and the courts, may be inclined to make this judgement for a wide range of polluting activites, it must be recognized that in some cases mitigating circumstances may suggest that justice be tempered with a little mercy.[23] The problem is to define the relevant mitigating circumstances, in particular to identify the circumstances in which the polluter could reasonably be said not to be responsible for the pollution damage that its pollution creates. In other words, how strict should the polluter liability be? In practice the common law recognizes many potentially mitigating circumstances in the pollution cases, typically involving only one pollutee, which come before the courts as actions in private nuisance.[24] Centralized instruments such as regulation and charges are inevitably less sensitive to the particular circumstances surrounding polluting activities. But a number of observers have argued that use should be made of subsidies to polluters, instead of regulation or charges, where the damage resulting from a pollutant increases owing to changes in the polluters' environment that are beyond their control.

The question that has exercised economists is whether polluter subsidies *could* be socially efficient, if well designed, in the way that charges would be in the idealized world.[25] The answer, in brief, is that the polluter subsidies could be socially efficient in the short run, but they are likely to be inefficient in the long run. Let us consider the short-run argument now and leave the long-run issues to section 4.2.5.

Suppose that the pollution control agency knows that each polluter, if left to its own devices, will pollute to the level E_n^A associated with q_n^A in figure 4.1 (b).[26] If the agency states that it will pay the polluter a subsidy of wv per unit of pollution abatement below E_n^A, it is apparent that by polluting at all the polluter will forego the amount wv on each unit of pollution. The subsidy wv per unit therefore represents the cost of polluting just as a charge of wv per unit would. Consequently, the individual polluter will adjust to the socially efficient level of its output q_2^A under either policy instrument.

The only difference between the two policies is that, compared with the no-policy situation, the firm's maximum profit level is raised by the subsidy but reduced by the charge, as the profit curves in quadrant (c) of figure 4.1. show. The difference in profit levels has no implications for resource allocation *in the short run*, so the conclusion is that in their idealized form the two instruments are equally socially efficient in the short run. However, even this qualified conclusion on the equivalence of the two instruments ignores the problem of fund-raising for the subsidy. If funds are raised by taxes, such as income taxes, which affect resource allocation in other respects (for example, by altering the choice between work and leisure), there will be other possible sources of inefficiency to be placed on the debit side of the subsidy's balance sheet. In addition the subsidy does nothing to help the just treatment of pollution victims since it concentrates its attention on the just treatment of those who are generating the pollution.

4.2.5.   *Long-run effects of the policies*

We have seen that in the idealized world of good information the pollution control agency can induce each polluter to limit its pollution to the socially efficient level by a fixed rate charge, by regulation or by subsidy. However, the maximum profit levels attainable by the polluter under each policy are ranked π (subsidy) $> \pi$ (regulation) $> \pi$ (charge) as in figure 4.1 (c). This will lead, in the long run, to a differential rate of entry to, or exit from, the competitive polluting industry, so that the *industry* output (and pollution) levels will eventually be ranked Q (subsidy) $> Q$ (regulation) $> Q$ (charge). Which of these is the socially efficient long-run industry output (pollution) level? The answer is that none of them may be, because the regulation, subsidy and fixed rate charge do not *exactly* internalize the pollution costs.[27] This can be seen in figure 4.2. The regulation fails to make polluters bear the pollution cost up to E_2 (i.e. *Ovw*); the subsidy not only fails to make the polluters bear *Ovw* but also offers a cost reduction of *vsz* on the pollution above E_2; the fixed rate charge, on the other hand, *over*-charges the polluter because the revenue *Olvw* exceeds the pollution cost by *Olv*. The socially efficient number of firms will enter the industry if and only if the profit level is that resulting from the exact internalization of pollution costs. This profit level will lead to the socially efficient size of the industry and the socially efficient industry pollution level. Compared with this, the industry pollution level in the long run is too large under regulation, and *a fortiori* too large under the subsidy; but under the fixed rate charge it is too small. Which of these three do we prefer, given that perfect internalization is hard to achieve because it requires each polluter to face a tarriff of charges with a different charge for each extra unit of his pollution? This will depend on whether greater importance is attached to avoiding having some *pollution abatement* for which in the aggregate $MAC > MPC$ (as will happen under the charge), or to avoiding having some *pollution* for which $MPC > MAC$ (as with regulation and subsidy).

In reality the long-run industry size effect of the policies may be less important than the predicted impact of policies on individual polluters, for a variety of reasons. One reason is that socially efficient entry may be obstructed anyway by market imperfections; and another is that the degree of uncertainty attached to the MAC and MPC curves may be so great that we are reduced to cruder attempts to move the system away from a clearly socially inefficient, no-policy state towards some more or less arbitrary, but lower, target pollution levels. If policies are aimed at reducing polluters to individual pollution levels which may be above or below the unknown socially efficient levels (see section 4.3.2), then little can be said of the social efficiency of the industry sizes that are produced by the alternative policies for bringing about this move in the twilight by each polluter.

We are now, however, edging into the analysis of a less perfect world, and it is high time a greater degree of reality was introduced by examining the problems raised by our lack of knowledge of the present and the future.

4.3   POLLUTION CONTROL POLICIES WITH IMPERFECT INFORMATION

4.3.1   *Lack of information and uncertainty*

Any policy that aimed at encouraging a move to socially efficient pollution levels would need a good deal of information on the marginal and total abatement cost and pollution cost curves. It is not enough to know the abatement and pollution cost effects of a *slight* variation from the prevailing pollution level. The range of pollution for which cost estimates are required must encompass both the prevailing and the socially efficient levels. Yet neither of the cost functions is easy to estimate, although the estimation of abatement cost functions is more likely to respond to intensive research than is that of pollution cost functions. With aggregate abatement costs at least the costs are in general potentially measurable, whether the abatement consists of a change of technology, a cut in output, or moving to a new location. Often the required information may be largely in the hands of polluters, and outsiders have to rely on average estimates for a typical polluter and to gross them up for the aggregate cost estimate. This leaves unsolved the problem of finding out the degree of variation of abatement costs across polluters, a piece of information that is important in evaluating the merits of charges as against regulation (see section 4.3.2 below). Even pollution control agencies with powers to extract this information rely on the co-oper-ation of polluters in some degree. Not surprisingly, there-fore, the current stock of abatement cost estimates is extremely patchy, which leaves polluters in a strong posi-tion to resist control policies on the grounds that they would impose catastrophic costs on the abaters.[28]

If our knowledge of abatement costs for the multiple types and sources of pollutants is sketchy, the currently available information on the value of pollution costs is almost non-existent for many pollutants. The valuation of pollution costs requires both information on the magni-tude of the damage measured in physical units (tons of wheat destroyed, lives lost, degree of physical incapacity endured) and an agreed means of converting these into a common unit of measurement, money values. The former type of information is easiest to collect for large-scale pollution occurrences such as oil spills which have an identifiable impact on the proximate environment and people and firms therein. For those residuals that are transported large distances, and residuals whose impact is not scientifically established (such as recently developed inorganic chemicals), and in cases where there is a risk of interaction between different residuals, even the estima-tion of damage in physical terms may be problematic.[29] Valuation problems are most easily resolved for the damage to property and production where market prices are available; they are least tractable for damage involv-ing significant psychological elements, such as the loss of pleasure owing to the destruction of amenities (by noise, smell and visual pollution) and pain and suffering. Attempts to find general measures of the value of damage owing to air pollution and noise from the market responses of pollutees have studied the relationship between pollution levels and house prices. But this approach is the subject of a current controversy and the results cannot be regarded as established knowledge at the moment.[30]

So far in the discussion of damage costs we have concentrated attention of the difficulties of measuring the value of pollution damage once it has occurred. An additional problem relates to the fact that at the point of time when a polluter is selecting the level of his activity that generates pollution, he may not know:

(a) how much pollution it will generate when emissions are an accidental by-product of the polluting activity; and

(b) how much physical damage is likely to result from his emissions of a given size, owing to the fact that other variables, outside his control, determine the amount of damage that is produced by a given quantity of pollutant emitted at any point of time.

Thus an oil-shipping company may find it hard to predict the incidence of serious tanker collisions at sea involving its fleet, and also the degree of environmental damage per barrel of oil spilled, which will depend on the location of the accident, the wind direction and state of the tide, etc. The result of this uncertainty about the future makes it difficult for the policy-makers to identify, at the time when the control policy has to be implemented, the level of the polluting activity that will be socially efficient.

This accumulation of obstacles to the identification of socially efficient levels of pollution inevitably leads one to the conclusion that the immediate attainment of social efficiency is a pipedream in the current state of knowledge. However, our theory does tell us the information we need in order to attempt to bring about socially efficient pollution, and we can hope to move towards this utopia as our understanding of pollution impact improves. The fact that perfection is currently unattainable should not be interpreted as a case against trying to improve the state of the world created by uninhibited polluting behaviour.

4.3.2 *Policies using poor information*

There are two ways of attempting to overcome the lack of information while implementing control policies:

(1) pursue the quest for socially efficient solutions through iterative control (or improvement little-by-little);

(2) use control policies to move the system towards pollution limits that do not claim to be socially efficient (let us call them 'emissions standards' or simply 'standards').

The use of iterative procedures and the pursuit of environmental standards are not in fact mutually exclusive, since one might very well adjust standards as part of an iteration towards stricter controls. However, when dealing with method (1) above we shall consider iteration as a means of achieving social efficiency, while method (2) explicitly abandons the attempt to achieve social efficiency. Let us consider the two methods in turn.

(1) *Iteration*

An iteration is a series of small steps designed to replace a larger single change, in the belief that little-by-little reduces the risks of missing the target, namely the socially efficient pollution level. There is a certain appeal about this proposal to the risk-averse policy-maker. However, it does involve a number of snags, not least of which are the assumptions that we may not know what the promised land looks like, but we shall recognize it when we get there, and correspondingly that we shall know, once we have taken it, whether any of the steps on the way was worthwhile or should be retraced. With these assumptions accepted it would be easy to believe that the iterative process would converge on the socially efficient pollution level. But the only information that an iterative process of pollution reductions is likely to generate is how much the (ambient) quality of the environment improves, in terms of the quantities of residual therein, for given changes in the level of pollution abatement required of individual polluters. This is not sufficient to identify the amount of damage produced by the residuals even in physical terms (deterioration of health, etc.), let alone in terms of the value people place on the damage. Consequently, although the iterative process would generate some information, it would not be sufficient to guarantee an approach to social efficiency even if the marginal pollution cost curve were upward-sloping as is conventionally assumed.

At least, in this conventional situation, if the guesstimates of the abatement and pollution costs were to indicate that, for each abatement step, marginal pollution cost exceeded marginal abatement cost, then we should know we were heading towards social efficiency. However, if the marginal pollution cost curve could be downward-sloping, the fact that for any abatement step the marginal pollution cost fell short of marginal abatement cost would *not* tell us that the step was socially inefficient. Referring back to figure 3.3 (e) on p. 66 it is apparent that each of the steps from q_s^A down to q_r^A has this characteristic, yet if the socially efficient output q_0^A is to be reached these abatement steps would have to be taken. Clearly an iterative procedure is myopic, judging each step in isolation, and this can give a very misleading view of the overall picture.

In addition to its inability to solve the deficiency of information, an iterative process may be administratively costly since it requires a succession of legislative and administrative changes. This applies to iterations involving changes in regulated standards or changes in charges.

Another problem is the risk of polluters responding to the early steps in the iteration in a way that proves costly when further responses become worthwhile in the subsequent steps. Thus, if a river polluter introduces a new process which reduces the concentration of a chemical in its effluent to ten parts per million, and later finds the river authority setting a regulated limit of five parts per million, a further adjustment of technology may be needed, perhaps even involving the scrapping of the first new process. This 'lock-in' to an initial abatement process may be mitigated by the adoption of more flexible abatement technology; but flexibility is not costless and switching costs could have been lower if the objective had been five parts per million at the outset. It is apparent, at least, that the evaluation of the benefits of little-by-little compared with larger changes is itself information-intensive, since it requires a knowledge of the sensitivity of the technology requirements to the precise pollution abatement target set.

Finally, there is a problem that is peculiar to iterations with a charging scheme. One of the claims that is sometimes made for charges is that they generate information on the polluters' marginal abatement cost curve. Assume that polluters are profit-maximizers, are short-sighted and do not collude. Then if the pollution control agency decides on the pollution level it wishes to induce, given *its* estimate of the marginal abatement cost curve, and then sets the appropriate charge, it will be able to observe whether polluters over- or under-react. If, for example, they reduce pollution by less than was estimated, this implies that marginal abatement cost is higher than estimated and the agency can revise its pollution target accordingly. However, if individual polluters deliberately under-react, in the sense of abating less than would maximize profits at the initial charge level, and particularly if polluters collude and under-react to a similar extent, then the response to the initial charge will not provide a signal that would lead the iteration process to converge on the socially efficient pollution level.[31]

It does not seem, therefore, that policy iterations are likely to prove a reliable means of reaching a socially efficient solution to pollution problems. This is not to say that little-by-little has nothing to offer, as we shall see in the next section, only that it is not going to be an infallible guide to utopia.

(2) *Standards-orientated control policies*

We shall now assume that the policy-maker has identified, for each pollutant in a locality, an appropriate emission standard, that is an appropriate amount of pollution in terms of quantity and concentration. 'Appropriate' in this context means that the total abatement cost incurred in satisfying the standard is expected, on the basis of best guesses, to be less than the benefit to society of the reduction in total pollution cost. The central question to be considered is whether information deficiencies introduce considerations that could influence the choice between the alternative control policies.

The policy analysis will be simplified at this point by not considering polluter subsidies further. The characteristics of this instrument in an idealized world of good information have been noted (section 4.2.4), and information gaps do not raise any problems for the instrument that are different from those that will confront a charge. Also, zoning will not be discussed in this section so that we can concentrate our attention on the much-debated choice between charges and regulation. The question that we seek an answer to is: what are the conditions under which we might favour a charge scheme, and what are the conditions under which regulation would be preferable? It should be borne in mind throughout that we are concerned with sub-optimization, that is with the minimization of the costs of achieving the pre-determined standards which themselves represent more or less arbitrary targets.

Perhaps the most frequently offered view of the relative efficiencies of charges and regulation is that charges are superior because they minimize abatement costs.[32] More exactly, the argument is that, if, in a locality, an overall emissions standard is set as the target for polluters as a group, then in response to a charge the required abatement by the group will be shared between polluters in the least costly way. Moreover, this will be achieved without the policy-maker knowing individual polluters' marginal abatement cost curves. On the other hand, it is argued, a policy of regulation could achieve this least-cost allocation only if the individual polluters' abatement costs were revealed to the policy-maker. Clearly, in a realistic world of costly (or simply missing) information this could be an important advantage for the types of pollutants that are emitted by numerous and heterogeneous sources.

We have seen before (pp. 108–10 in connection with figure 4.3) that the total abatement cost incurred in satisfying an overall group standard ($E_{\frac{1}{2}}^{A+C}$ in quadrant (c)[33]) is minimized by a charge Ot. A similar outcome under regulation requires $E_{\frac{1}{2}}^{A}$ and $E_{\frac{1}{2}}^{C}$ to be known and imposed as individual standards. If instead a *uniform* individual standard of $\frac{1}{2}E_{\frac{1}{2}}^{A+C}$ were imposed on each polluter by regulation, the total abatement costs for polluters A and C together would be raised as we saw on p. 109.

Whether, for any given pollutant, the cost saving with a uniform charge, relative to regulated uniform individual standards, would be great depends on the degree of difference between the two cost curves MAC^{A} and MAC^{C}. If the two curves were the same, then uniform individual standards would also minimize abatement costs, by imposing the same pollution levels that the polluters would choose under a charge. The regulatory authority would share out the overall pollution limit equally among the individual polluters. If on the other hand the MAC^A and MAC^C curves were widely divergent, then the uniform charge *would* achieve abatement cost savings. It is hard to generalize about the quantitative significance of the cost savings given that different pollutants are generated in quite different ways, some arising from many very similar sources (such as fumes from car exhausts, soot from domestic chimneys, sewage from households and nitrogen from farm run-off) and others from a mixture of sources (such as organic and inorganic chemicals and metals discharged to rivers from many heterogeneous industrial activities). For some pollutants the cost savings may be significant, for others not. In their eagerness to justify charges schemes economists are sometimes guilty of a double-think as far as the cost savings are concerned. On the one hand they claim that, in general, abatement costs do vary widely and that this implies substantial cost savings. On the other hand they justify the pursuit of environmental standards on the grounds that total abatement costs are often quite small.[34] They really cannot have their cake and eat it! If the total abatement costs for the group are small compared with other production costs, then the relative advantage of a charge scheme is going to be small as well. The evidence that is usually quoted for expecting large cost saving stems from a number of studies of river basins in the United States.[35] Unfortunately, the evidence quoted is not sufficient to judge the significance of the cost savings. If we are told that total abatement costs will be, say 50 per cent lower under a uniform charge than under uniform individual standards, we do not know whether that implies a large *absolute* abatement cost saving unless the absolute figures are given. Even if the absolute savings are quoted we still need to know their size relative to the other production costs of the polluters, so that we can estimate the impact of pollution control on prices and profits, employment and so on under the two policies. Only then can we judge the significance of the cost savings under a charge in terms of the likely response of polluters.[36] Certainly we should not meekly accept the synthetic fact that the cost savings would usually be very significant indeed.[37] In some cases they might, in others they might not.

The significance of the potential abatement cost savings will eventually be determined through numerous empirical studies of abatement costs for particular pollutants. But we can now consider another, fundamental theoretical question. Is the minimization of total abatement cost desirable? The answer is yes, as long as units of pollution from the many sources would cause equal damage once they are added to a given total level of existing pollution. This requirement is implicit if we regard as desirable an allocation of a pollution abatement obligation between difference sources that is based on abatement costs alone.

Once it is recognized that the impact that any unit of pollution has on society will depend on the location of the polluter (so that the marginal damage curves of polluters differ), the control agency, even without accurate estimates of pollution costs, may not be satisfied to set an overall group standard and leave the allocation of pollution units to the free responses of polluters.[38] An attempt could be made, for example, to set individual emissions standards that vary between polluters (or between groups of polluters) according to location; for example, stricter standards on river polluters who are upstream than on those near the river mouth, and stricter standards on air polluters who are adjacent to residential areas than on those further away.

Consider the situation depicted in figure 4.4. The pollution control agency decides that an appropriate overall standard for the group of polluters (A + C) on a river is represented by the aggregate pollution level $\bar{E}^{A+C}$ in

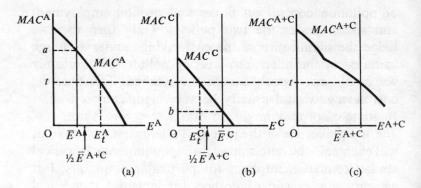

FIGURE 4.4   *Charges and regulation when pollution from different sources has different consequences: (a) polluter A: upstream, (b) polluter C: downstream, (c) total A + C*

quadrant (c). Now the impression formed by the agency is that the effluent emitted by the upstream polluter, A, is approximately twice as harmful (in terms of residuals concentrations in the river and implied pollution costs) as that of the downstream polluter, C. It is therefore inclined to set twice as stringent a standard ($\bar{E}^A$) for polluter A as that ($\bar{E}^C$) for polluter C. In other words, the agency's preferred position, on the basis of environmental impact, is for $\bar{E}^{A+C}$ to be emitted in total, with $\bar{E}^A$ entering the river upstream and $\bar{E}^C$ entering downstream. However, the agency wisely consults an economist from Environment for the Future, and is correctly shown that using the regulatory approach to enforce the individual standards $\bar{E}^A$ and $\bar{E}^C$ would fail to minimize the sum of the two polluters' abatement costs. This is apparent from the fact that the last unit of abatement before $\bar{E}^A$ is reached by A is very much more costly than the first unit of abatement below $\bar{E}^C$ which polluter C naturally does not undertake. The economic adviser, keen as ever to spread the doctrine of pricing infallibility, urges the adoption of a charge equal to Ot. This, he says, will allocate the aggregate pollution $\bar{E}^{A+C}$ in an efficient manner, as the invisible hand is wont to do. Mindful of the irrefutability of the doctrine of infallibility, the agency introduces the charge and the market allocation of the aggregate pollution proves to be E_i^A, E_i^C, the reverse of the allocation the agency had originally set its heart on $\bar{E}^A$, $\bar{E}^C$! In fact, the market allocation is even further away from $\bar{E}^A$, $\bar{E}^C$ than a uniform standard of $\frac{1}{2}\bar{E}^{A+C}$ would be, as figure 4.4 shows.

This parable of the wiser adviser points to the dilemma of a choice between two simple policies. The regulatory approach of differentiated individual standards, selected on the basis of environmental impact with no consideration of abatement costs, proves to be an expensive means of reaching the overall standard $\bar{E}^{A+C}$. On the other hand reducing total abatement costs by implementing a uniform-charge scheme means that $\bar{E}^{A+C}$ generates more damage than under regulated individual standards, because the market allocation of pollution shares ignores the difference in environmental impact between E^A and E^C. We find, therefore, that the pollution control agency faces the problem of needing at least some approximation of the abatement cost savings that a uniform charge would yield, in order to judge the merits of allowing the greater environmental impact by not discriminating against the upstream polluter, an impact that itself is likely to be to some extent impressionistic. It is not enough simply to follow the advice that economists frequently give, namely to set a uniform charge.

It is in recognition of this realistic problem that some economists have explored the possibility of implementing differentiated rates of charge which take account of the differences in the environmental impact of various polluters' effluent.[39] If the whole group of polluters were divisible into an n-sized sub-group of A-types, who all have the same abatement cost curve MAC^A in figure 4.4, and an n-sized group of C-types, whose abatement cost curves are like MAC^C, then a differentiated charge has similar consequences to individual standards.[40] If the charge levels are set at Oa for each polluter of the A-type and Ob for each polluter of the C-type, in an effort to induce the two groups to keep within the overall limits of $\bar{E}^A \times n$ and $\bar{E}^C \times n$, then the differential inducement to abate violates the requirement of abatement cost minimization. This requirement is that each polluter abates to such an extent that the level of marginal abatement cost reached is the same for each (equal to Ot under the uniform charge). But with the differential charge each polluter sets his marginal abatement cost equal to his *own* rate of charge, and the marginal abatement costs of the two groups will differ. The implication is that, in this case, using the differential charge achieves the ratio of pollution shares $\bar{E}^A/\bar{E}^C$ originally desired by the control agency, only at the expense of abandoning abatement cost minimization.

Where a differential charge *would* have an advantage over differential individual standards is the situation in which abatement costs differ between polluters *within* the A group or the C group. If, for example, the upstream A group is a heterogeneous collection of n polluters, then the charge Oa which is uniform for that group allows different levels of abatement *between* the n members of that group. Taken together, however, the A group abates to the required level of pollution $\bar{E}^A \times n$. Suppose, instead, that a regulatory approach were used by which the two groups' overall standards, $\bar{E}^A \times n$ and $\bar{E}^C \times n$, were differentiated, but uniform individual standards were imposed within each group; then each A-type polluter would have to restrict its pollution to its share of the standard, i.e. to $\bar{E}^A$. Consequently the total abatement cost of the A group would be higher under regulation than under the charge. However, it is clear that the case for a charge in place of regulation now rests on the ability of the pollution control agency to identify polluter groups between which there are differences in environmental impact and within which there are significant variations in abatement costs. This identification requires considerable empirical information. The economic theory therefore tells us which data are relevant to a resolution of the problem of instrument choice; it does not in itself offer an adequate basis for making the choice.

To round off the discussion of charges and regulation there are three other possible implications of the choice of policy worth mentioning. These relate to:

(a)   the consequences of the incorrect estimation of the abatement cost curve (assuming for simplicity that marginal pollution costs are known);
(b)   the size of administration and enforcement costs;
(c)   the incentive of polluters to develop 'clean' technology.

(a)   In figure 4.5 the actual, true marginal abatement cost curve lies above the estimated curve. Basing its judgement on the estimated curve, the pollution control agency aims to achieve the pollution level $\bar{E}$. If it sets a regulated standard at $\bar{E}$ then the policy involves the social loss *abc* compared with the socially efficient situation at E_s. If, instead, the agency sets the charge at *Ot*, equal to the marginal pollution cost at the *estimated* socially efficient pollution level $\bar{E}$, the profit-maximizing polluter will in fact pollute to E_p. The relative social loss here is clearly *dce*.[41] Whether the loss under regulation exceeds the loss under the charge (relative to the socially efficient state) depends on the slope of the marginal pollution cost curve. If it is steep, so that the error in an upward direction in the pollution level under a charge ($E_p > E_s$) incurs heavy extra pollution costs, then the cautious solu-

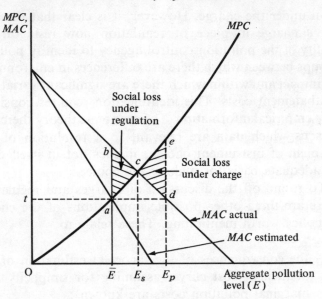

FIGURE 4.5   *Charges and regulation when MAC is incorrectly estimated*

tion under regulation ($\bar{E} < E_s$) avoids the heavier relative social loss. However, this does not really provide a basis for preferring regulation unless we *know* that *MAC* is under-estimated (in which case why not alter the estimate?). For if *MAC* could equally well be *over*-estimated the regulatory approach would lead to a pollution level in excess of the true socially efficient level, while a high level of charge chosen on the basis of the exaggerated abatement cost would lead to pollution below the socially efficient level. The situation is now reversed and the mistake in estimating *MAC* leads to a low pollution level under a charge which is preferable to the higher level under regulation if the *MPC* curve is steep. (The reader can confirm this by redrawing figure 4.5 with the *MAC*-actual and *MAC*-estimated curves reversed, and the charge set where the *MAC*-estimated and *MPC* curves intersect.)

(b) One of the favourite arguments of the advocates of charges is that the instrument is superior to regulation because it is cheaper to administer and to enforce.[42] This is another issue that is difficult to resolve entirely on theoretical grounds, but a few general propositions can be made.

Let us distinguish between two types of regulation, the uniform individual standard and the set of individual standards that allows diversity of pollution abatement, and consider the requirements of information, administration and enforcement in turn.

A uniform standard regulation involves no cost of information. The target standard is known, and this is embodied in a regulated ceiling on pollution. But to induce firms to meet this standard through a charge requires either extensive information on abatement cost functions internal to the firm or an iterative procedure to approximate the charge that induces the required overall reaction. To minimize the lock-in effect previously referred to and to avoid the uncertainty to firms inherent in an unconstrained iterative procedure, some information on likely reactions is necessary, so that the information costs must be higher than for a uniform-standard regulation.

Administrative costs of an instrument may include the monitoring of pollution levels and the running of the facilities for making the payments (in the firm and in the government department). In fact regulation requires only the former, while a charge requires both because it involves a transfer of funds. It is important not to be misled by suggestions that charges have the advantage of being self-financing. We are talking here about real resource cost, and the fact that the government receives funds with a charge does not reduce the *total* cost of the instrument, but rather the reverse, since the transfer of funds has to be administered and the use of the funds supervised.

As regards the cost of enforcement, regulation involves the punishment of firms that allow their pollution levels to exceed the regulated standard, either through a set of automatic penalties imposed by the government department or through litigation in the courts. These costs could be high, but will they be any lower with a charge? It seems certain that they will be no lower, since the pollution ceiling must be enforced regardless of the instrument used, and a charge introduces an extra dimension for evasion through non-payment of the charge made for the level of pollution reached. A charge introduces a multitude of tailor-made standards for individual firms, depending on the payment made, and observing a 'high' level of emissions from a firm is not sufficient for action to be taken: a higher level of emissions than the one that the firm has paid for must be observed, which also complicates the required monitoring machinery.

These arguments suggest that the differential in information, administration and enforcement costs tend to favour the regulation of uniform standards rather than charges. But a regulated uniform standard may involve higher costs of pollution abatement than does a charge, so let us compare charges with a regulated *set* of individual standards which allows diversity in pollution levels between firms.

The set of standards obliges firms to demonstrate exceptional abatement costs if they are to be allowed to deviate from some emission norm. While this does not involve the government in the costs of extracting information, it does require a system of evaluating the firms' estimates. Consequently the information costs will be higher than with uniform standards, and probably they will be higher than for charges under the iterative procedure. However, it is likely that they will be no higher than under a charge without an iterative procedure which requires the extraction of information on abatement costs in order to set the charge.

The administrative costs of a set of standards are likely to be higher than for a uniform standard, though again they do not involve the resource cost of a financial organization. The monitoring may be more complex when no uniform criterion of excessive pollution is applicable to firms, but a charge involves a similar degree of complexity and the financial machinery as well. The tentative conclusion must be that a charge is as administratively costly as a set of regulated standards unless an iterative procedure can be used for the charge.

The costs of enforcement through the punishment of delinquent polluters need be no higher for a regulated set of standards than for a regulated uniform standard so that given the previous reasoning on the relative costs of enforcing a charge and uniform standard regulation, we can say *a fortiori* that a charge will be no less costly in this respect.

The one clear conclusion that emerges from this survey of information, administration and enforcement costs is that the regulation of a uniform standard is almost certainly cheaper to operate than a charge. On the other hand the balance of these costs for charges and a regulated diverse set of standards cannot be determined using qualitative theory alone. But there can be no presumption that a charge is less costly. This conclusion is strengthened if we recognize that a uniform charge, which has been assumed in this discussion of administration and enforcement, may need to be abandoned in favour of a more complicated differential charge scheme owing to the diversity of environmental impact of pollution from different sources

(c) It has been argued by some economists that charges provide the polluter with a greater incentive to adopt known less polluting technology, and to develop new

ones, than does regulation.[43] This is a difficult proposition
to analyse fully, but two main points can be suggested.
First, there is no clear reason for expecting the choice
between *existing* clean and dirty production processes
to be different under a charge and a regulation which
aims to restrain the polluter to the same pollution level.[44]
Referring back to figure 3.4 on page 70, as the polluting
firm abates from E_n^A to E_j^A it switches from process $P1$
to $P2$ in quadrant (b). But this switch of process would
be undertaken whatever the method by which the pollution
control agency induces the move to E_j^A.

Second, to make predictions about the degree of innova-
tive zeal under the two policies we need to have a theory
of the determinants of innovation. This has not been
offered by those who claim the superior incentive under
a charge. Their view is based on the fact that profits
are squeezed more severely by a charge than by regulation,
owing to the burden imposed by a charge on the units
of pollution below the standard. If declining profits were
generally associated with rapid innovation this argument
would be persuasive. But much innovation is thought
to relate to a high rate of investment in new machinery,
and this is unlikely to be characteristic of profit-squeezed
firms.[45] However, the profit-squeeze hypothesis of innova-
tive behaviour is empirically testable, so that those who
press the hypothesis in relation to charges will perhaps
produce the evidence in the future. The currently available
empirical knowledge of innovation does not appear to
provide supporting evidence, since there has been little
apparent relationship between monopoly power (and
therefore, presumably, profitability) and innovation.[46]

4.3.3   *Policies under uncertainty about the future*

(1)   *Stochastic influences*

Pollution control agencies are confronted not only with imperfect information about the past and present impact

of pollution; they also face the problem of forecasting the impact of pollutants on an environment whose ability to assimilate them, and thereby to moderate the damage, varies through time. The assimilative capacity of the environment varies both seasonally and within a season, according to climatic variations. The amount of rainfall affects the ability of rivers to dilute effluent, and the atmospheric conditions determine the rate of dispersion of air-borne pollutants. The problem is that the agency does not know in advance which state of the environment will prevail. In effect the marginal pollution (damage) cost curve will change position with the circumstances. Over a period of time the position of the curve on different 'days' will be randomly distributed across a range of experience.[47] To simplify the argument let us assume that there are only two possible states of the environment, normal conditions and crisis conditions, characterized respectively by normal and poor assimilative capacity.[48] Under normal conditions the marginal pollution cost for polluters as a group would be MPC (normal) in figure 4.6, whereas if crisis conditions occurred MPC (crisis) would be experienced. The question is which level of charge or of regulated standard should be set for social efficiency if the MPC curve fluctuates randomly between the two positions? For it is clear that, with marginal abatement cost as shown in figure 4.6, the socially efficient pollution level may turn out to be either E_1 or E_2 on a particular 'day'; and if on any 'day' the wrong one of these two levels is chosen the result will be a social loss. We can now describe two alternative possible scenarios to clarify the socially efficient pollution control strategy.

The first scenario is one in which the pollution control agency knows *ex ante* the probability of the next 'day' being normal or a crisis, but having chosen the policy *ex ante* it cannot alter it during that 'day'. Now if either a charge or a regulated standard is used to limit pollution

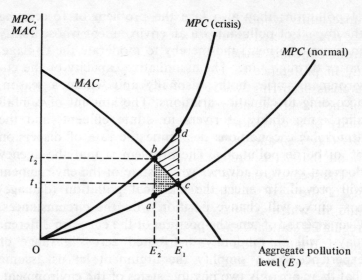

FIGURE 4.6   *Charges, regulation and random influences on pollution costs*

to E_1 in figure 4.6 then this will be socially efficient if and only if the marginal pollution cost curve happens to be *MPC* (normal) on the 'day'.[49] If instead the 'day' turned out to be a crisis-day then the social loss would be *bcd* (the excess of *MPC* (crisis) over *MAC*) as compared with the socially efficient pollution level E_2. Choosing E_1 therefore incurs losses if a crisis occurs. If, on the other hand, the control agency were very risk-averse and chose *ex ante* the pollution level that would be socially efficient in the event of a crisis, then again a social loss could occur. With the pollution target E_2 established, if the 'day' proves to be quite normal, so that *MPC* (normal) prevails, then there is a degree of 'overkill' in the control policy. The area *abc* represents the social loss associated with E_2 relative to the selection of the socially efficient pollution level E_1.

A rational approach for the agency in the light of this

uncertainty would be for it to attempt to maximize the *ex ante* expected value of the gains to pollution control from choosing either E_1 or E_2. Thus it would choose E_1 in preference to E_2 if *bcd* times the probability of a crisis occurring was less than *abc* times the probability of the 'day' being normal. Therefore the greater the probability of a crisis and the greater the difference between *MPC* (crisis) and *MPC* (normal), the greater is the likelihood that E_2 is preferable to E_1 on efficiency grounds.

The second scenario is one in which the *ex ante* probability of the next 'day' being normal or a crisis is known, and in which it is possible to tighten up the level of regulated standard early in the day when a crisis is recognized. Baumol and Oates use this situation as a justification for using 'direct', regulatory, controls as a means of handling short-run environmental crises.[50] They claim that charges are administratively hard to alter at short notice, and that even if some adminsitrative flexibility could be devised charges are unsuitable for controlling rapid variations in environmental impact because polluters only respond to charges with a substantial time lag. The implication of this argument is that if a charge is used to control pollution then the agency is committed to its *ex ante* choice between E_1 (using a charge of Ot_1) and E_2 (using OT_2). The balance of losses for the wrong choice of E_1 or of E_2 is the same as in the first scenario. However, if a regulated standard is used, the presumption is that the *ex ante* decision to select E_1 can rapidly be changed in favour of the tighter standard E_2 when the warning signs of an environmental crisis become apparent early in the day. At the extreme this flexible standard achieves the socially efficient level of pollution whether the day proves to be normal or a crisis, but even viewed more moderately the greater degree of flexibility and higher speed of polluter response that the regulatory policy offers may provide a means of reducing the social losses (which result from incorrect forecasts of MPC) compared with a charge. This further consideration in the regulation versus charges controversy simply serves to confirm the view that the case for abandoning the regulatory approach, favoured by some economists, must ultimately rest on the production of empirical estimates of the costs and benefits attached to each policy instrument in specific pollution situations. Baumol and Oates favour using a charge to induce polluters to restrict pollution to E_1 in normal circumstances, together with a regulated standard such as E_2 to handle crises.[51] This policy mix might be more efficient than a pure-regulatory approach if substantial abatement cost savings were yielded by the charge on normal 'days'.

(2) *Accidental pollution*

Some of the most spectacular cases of pollution are the result of accidents occurring within production operations that, when operating correctly, do not pollute at all or at least not as intensively. Explosions at chemical factories, accidental oil spills from land-based sources (e.g. refineries) and from tanker groundings and collisions, fires at depots storing toxic chemicals, emissions of smoke and chemicals to the air owing to boiler failure – these and many others are to some extent accidental in the sense that the pollution is not intended. In such cases social efficiency considerations are relevant at two levels:
(a)   the identification and achievement of a socially efficient level of risky activities;
(b)   the identification and achievement of an efficient balance between controlling the emission of a pollutant once an accident has happened, mitigating its impact on the environment, and accepting some degree of environmental damage.[52]

As far as level (a) is concerned the policy problem is particularly acute for those accidents that are infrequent and unpredictable but of great severity when they occur. The socially efficient incident of such accidents is difficult to identify *ex ante*, even with a careful appraisal of the currently available evidence by an objective observer (not the polluter!). Even where the frequency and predictability of accidents is greater, attempts to induce socially efficient levels and modes of operation of risky activities, and consequently socially efficient accident rates, face considerable difficulties. Where the accidents relate to activities operating within national boundaries it is possible to establish regulated 'due care' standards to control the polluter's risk-taking. Such standards typically involve a technology evaluation by the control agency, and a blueprint for acceptable operation of a risky undertaking. With potential damage costs unquantifiable in some respects, 'due care' standards inevitably involve the arbitrary element characteristic of the standards approach to pollution control. In addition the conventional objection to 'due care' standards is that they specify a particular form of operation. This may impose biases in technology selection, which can be socially inefficient if the agency is less informed than the polluter about the costs of alternative abatement methods. However, the potential social benefits from inducing 'due care' are large for some types of production (e.g. atomic energy plants and chemical firms involved in dangerous processes), and unless the cost to the polluter resulting from the restriction on his choice of abatement process is great the interference with the choice will be socially efficient. One of the limitations to its applicability, however, relates to the conduct of risky activities across national boundaries. The classic case is the shipment of oil by tanker. Here the tanker owners may be able to escape the requirements of the 'due care' laws legislated by particular nations (usually the ones polluted rather than polluting) by flying flags of convenience, and by operating within a system of hold-

ing companies which effectively limits the liability of the
parent company if the laws of a country are broken
and the violation is detected.[53]

The problem is that the policy instruments face obstacles
that may be equally, or even more, insurmountable. A
fixed rate charge can scarcely be used where accidents
are infrequent and unpredictable, since any single rate
of charge will be excessive during lengthy periods without
an accident and inadequate as a means of internalizing
the large external costs generated by the occasional accident. The difficulty is that the charge needs to be related
to the accident record of the operator of the risky activity
if there is to be an appropriate incentive to take 'due
care', but the absence of informative accident records
where accidents are infrequent would make the charge
rate arbitrary. In addition, pollution that relates to international production activities involves accidents by companies that are outside the offended nation's tax jurisdiction; this presents difficulties of enforcing the payment
of the charge.

A third possible policy is the requirement that those
involved in risky activities take out compulsory insurance
against liability for damage arising from accidental pollution events. If the insurance is allied with the strict liability
of polluters for damage incurred by pollution victims,
an efficient incentive to take care may be created. This
would require that the insurance premium be related to
the insured party's activity level and to his past accident
record. It is the difficulty of achieving accurately risk-related premiums that provides the obvious obstacle in the
infrequent, unpredictable accident cases. But the more
widespread the risk, as with worldwide tanker operations,
the more feasible is risk-related insurance. There is also
the additional difficulty of non-measurable damage costs.
Insurance companies do not like open-ended commit-
ments, and they would usually insist on ceilings being

placed on their liability to compensate for damage costs incurred through an accident by the insured. This prevents the correct incentive to the polluter to take care unless it is liable for any residual damage costs. This throws the onus on the court awards of damages against the polluter rather than against the insurance company, an obligation that the courts may find difficult to satisfy.[54] Having said all this, the possibility of combining compulsory polluter insurance and strict liability for damages remains worthy of greater exploration, not least because it offers compensation to victims together with a modicum of incentive to polluters to take 'due care' as long as some attempt is made to risk-relate the insurance premia.[55]

The second level at which efficiency considerations are relevant to accidental pollution concerns the post-accident response by the polluter and by the government with jurisdiction. Once an accident has happened an attempt can be made to control the emission (for example, the attempt to cap an oil well that has 'blown') or to control the resulting damage (for example, the use of chemical dispersants on oil slicks), or the damage costs can be accepted. In other words there is a post-accident choice between alternative combinations of abatement cost, damage reduction cost and damage cost. There is nothing analytically new for us in the choice between abatement in some form and suffering the damage costs, but in cases of accidental pollution it is often informative to make explicit the more complicated choice when abatement (pollution prevention) *and* damage reduction can be undertaken. Let us illustrate with the case of oil tanker accidents. It is normally the case that the government allows salvors the opportunity to save the ship. Yet lengthy salvage operations can, if they are unsuccessful, ultimately *raise* the cost of clean-up (damage reduction) and of environmental damage, as almost certainly was the case with the Torrey Canyon disaster.[56] One of the problems with the current salvage arrangement, the Standard Form of Salvage Agreement of Lloyds known as the 'no cure – no pay' contract, is that it very much favours the salving of the ship even at the expense of dumping some of its cargo if this cannot easily be trans-shipped owing to rough sea. This is because the hull of a large tanker is worth about ten times the value of its cargo. It is apparent that the salvage contract does not provide the correct incentive to salvors to minimize external costs; yet the subsequent costs of clean-up etc. to other parties may substantially exceed the value of the ship that the salvors are interested in saving. In addition the present first-come, first-served awarding of the salvage contract can accentuate the risk of environmental damage. In the Torrey Canyon case, for example, easily the largest tug available failed to win the contract and left the scene. These two problems with salvage contracts suggest two forms of government control. First, where the risk to the environment is great the private contract could be overruled and a greater incentive to save the cargo be introduced. Second, the government could pay a daily allowance to hold selected salvors in the vicinity to facili-tate subsequent sub-contracting from those who win the contract to those with more suitable tugs.

4.4 THE DISTRIBUTIVE EFFECTS OF POLLUTION CONTROL

When one considers the complexity of pollution in the real world it is not surprising, perhaps, that it is difficult on purely theoretical grounds to predict who will benefit and who will lose, on balance, from the investment of resources in pollution control. Is pollution control élitist in the sense of yielding net benefits that *as a proportion of income* are greater for the rich man than for the poor?[57]

In other words, is pollution control regressive in its distributive effect?

At the end of this section we shall outline the results of some of the empirical studies of the incidence across income groups of the benefits and costs of pollution control. But first a few general points can be made to throw light on the proposition that the distributive consequences of pollution control cannot be predicted from theory.[58]

Even if attention is limited to the *benefits* of pollution control it is hard to make a categorical prediction of the nature of the relationship (if any) between benefits received and income level. The reason is that there are two influences on the distribution of benefits which may conflict with each other. In the first place, if pollution levels are fairly uniform across a pollution control area then the benefits of abatement are a public good, equally available to all. If everyone values the pollution reduction equally the benefits will be spread evenly across the income scale and will consequently represent, as a percentage of income, a greater benefit to low-income recipients. But if the rich attach a greater value to a given amount of pollution reduction then this need not be the case: even the benefits of public goods can be progressive or regressive depending on the tastes of people in different income groups.

In the second place, to the extent that there are local variations both in the levels of pollution in the absence of control and in the degree of control exerted, the rich may benefit relatively by their ability to select a pollution-free place to live. Yet the fact that rich suburbs are polluted less than inner-city slums is not in itself evidence of a greater benefit of the rich from pollution control. It may simply reflect the access of the rich to the unpolluted areas which do not need pollution control and do not benefit from such policies. Only if the rich tended to live in areas that offer heavy benefits from control would there be a regressive element in the benefit distribution, to counter-balance any progressiveness that results from the publicness of control within areas. The regressiveness, progressiveness or otherwise of the benefit distribution can be found only by measuring the impact of particular pollution control schemes.

The distribution of the costs of pollution control is a complicating factor similarly hard to determine *a priori*. When a more tightly regulated standard is introduced, in order to comply with it the typical polluter will incur some extra abatement cost, the significance of which may vary between industries. If the standard is attained by the imposition of a charge then there is the extra cost to the polluter of the payment of the charge on remaining units of pollution. Whether the polluter abates by cutting output or by altering its technology, some increase in the price of the good produced is to be expected. The distributive effect of the price rise naturally depends on the characteristics of the good, in particular on whether it is purchased proportionately (to income) more by the rich or the poor. A full evaluation of the incidence of abatement costs therefore involves an investigation of the abatement costs associated with different product groups, and of the significance of different products in the consumption patterns of different income groups. Unfortunately the task does not end there because abatement can involve a loss of profit, together with a reduction in employment, particularly if output is cut rather than technology being altered. The loss of profit will be progressive to the extent that shareholders are relatively well off, but the employment effect is more likely to be regressive when firms react by laying off the least skilled members of their workforce.

The likely net consequences of all these changes on the benefit and cost sides are hard to judge, aren't they?[59] To round off this discussion some fragmentary pieces of evidence can be mentioned. Since we have observed earlier in this chapter that there is considerable uncertainty over the magnitude of pollution damage costs, it should come as no surprise to find that little is known of the distribution of the benefits of pollution control in the form of damage reduction. It is well known that the rich benefit more than the poor from the provision of national parks and other recreational facilities because they use them more.[60] In fact, the demand for such facilities is income-elastic, meaning that when people's real incomes rise by a certain percentage their demand increases by a larger percentage. But it does not follow from this that the rich necessarily benefit more from pollution control in the localities where people live and work. The rich may have greater access to distant recreational facilities, but if in general it is the poor who are most exposed to pollution risks at home and at work, then they may benefit more from pollution control, at least relative to their income and perhaps even in absolute terms. On the whole the patchy evidence available (for the United States) indicates, for example, that air quality is higher for the rich than for the poor, as measured by pollution exposure indices for different income groups.[61] But indices of exposure indicate the *quantity* of pollution to which people are subjected, not the *value* that people attach to a reduction in such exposure. Do people place a higher value on a given quantity of pollution abatement the higher their income levels? The fragmentary evidence does not appear to support the view that the value placed on abatement increases more than in proportion to income (i.e. that the income elasticity of demand for pollution control exceeds unity).[62] Taking the evidence on exposure indices and on income elasticities together we can say, very tentatively, that the existing evidence supports the proposition that the benefits (in terms of *value*) of pollution abatement do not increase more than proportionately as incomes rise. In other words, as a percentage of income the benefits may well fall as one moves up the income scale. At least there is no clear evidence that the benefits of pollution control are regressive (i.e., pro-rich) in their distributive effect.

This conclusion relates to the benefits of pollution control, but a complete evaluation of the distributive impact of such a policy should include the measurement of the consequences of abatement costs. On the whole the evidence on the incidence of abatement costs (again for the United States) suggests that they are regressive; as a percentage of income they are higher for the poor than for the rich.[63] The reason for this is partly that the poor buy more of the polluting products relative to their income, and partly that the cost of meeting the proposed standard for automobile emissions will impose a fixed cost per car that is higher, as a percentage of income, for the low-income earners than for the better off.

It should be apparent from this brief summary of the empirical evidence available to date that it is not possible to draw firm conclusions about the distributive effects of the *net* benefits of pollution control schemes currently proposed in the United States. Nevertheless, it is easy to see why resistance will be met to hard-hitting programmes that aim at tight standards, unless some attempt is made to moderate the impact of abatement costs on low-income earners, for example through polluter subsidies financed from general taxation.

CHAPTER 5

Pollution Control
and the Law

5.1 ECONOMICS, THE LAW AND POLLUTION CONTROL

As soon as one descends from the exalted plane of pure theory it becomes apparent that there is a discrepancy between the beautiful descriptive simplicity of policy measures in economic analysis, and the complexity of the legal apparatus designed for pollution control. An economist's first reaction is inclined to be that the law is over-elaborate and that his analysis goes to the heart of the matter, while a lawyer may initially feel that the economic models are simplistic, unrelated to a complex reality. The truth of the matter is that the real world *is* complex, and laws have to reflect the variety of motivations displayed by people and institutions and of circumstances in which they operate. And yet the abstraction involved in economic analysis does highlight the implications of different policy objectives and identify some of the advantages and disadvantages of different forms of legal control. The complementarity between the lawyers' and economists' views of the world would undoubtedly have been recognized earlier if, in English-speaking countries at least, the disciplines had not developed such separate identities.

This chapter does not aim to convert an economics audience into students of law! The intention is to provide

149

a brief introduction to ways in which the law operates in the context of pollution control, in the hope that the interested student of economics will read some of the references given to work by lawyers.

The involvement of the law in protecting people from polluting activities is not a new phenomenon. Instances of statutory intervention in Britain can be found at least as far back as Edward I's reign, when an Act was passed in 1273 to prohibit the use of coal which was thought to be injurious to health. In addition to such government intervention, for centuries individuals have had recourse to the private law to increase their freedom from uncompensated pollution damage. The involvement of the private law developed on a narrow base, relating to interference with the use and enjoyment of one's land. Consequently even to this day typical private nuisance actions are concerned with the harm caused by the activity of one land-owner to that of a neighbouring land-owner, and the private rights to protection from pollution damage are restricted to those with a legal interest in the occupation or enjoyment of land.[1]

The statutory intervention and private law actions together constitute a two-level system, with the private law component addressing itself to establishing private rights, and statutes in effect being introduced where gaps and deficiencies are observed in the protection offered by the *effective* operation of private law. Naturally statute law directs itself to broader environmental objectives than are usually relevant to the specific issues arising in a private nuisance conflict between neighbours. Let us consider briefly the kind of legal machinery that is available at each of the two levels of legal involvement.

5.2   PRIVATE LAW

Private law is concerned with the rights of individuals, and in theory these rights are not to be sacrificed to the interests of society as a whole. The central aim of a private law decision, again in theory at least, is the re-establishment of the state of affairs prevailing before the pollution damage was imposed, which is referred to by lawyers as the restoration of the *status quo ante*.[2] However, in reality there are a number of obstacles to the establishment of rights in favour of pollution victims (plaintiffs) in British courts. The plaintiff must have a legal interest in land; the interference must usually be more tangible than aesthetic harm (such as the deterioration of a view); the plaintiff's activity must not be unduly sensitive and so on.[3] If these conditions are satisfied the nuisance is actionable, and the plaintiff may seek remedy in court. The interesting question for the economist at this point is whether the court remedies differ significantly from the idealized fully negotiable pollutee rights that provided one starting point for the bargaining solution analysed in section 3.2 above.

The remedies available to the court when the polluter is deemed liable are a damages award and an injunction. A damages award is the imposition on the polluter of a liability to pay compensation to the victim at the level of damage as assessed by the court. An injunction, on the other hand, imposes a restriction on the polluting activity, usually that the polluter refrain from the continuation of the nuisance.

With idealized fully negotiable rights pollutees could release polluters from an obligation to eliminate pollution by accepting full compensation for damage, on a continuing basis for as long as the damage occurred. But there is some doubt as to whether the damages remedy provides such compensation. Quite apart from the courts' reluctance to compensate at all for non-tangible damage (e.g. the effects on mental health), preferring to relate damages calculations to the loss of the market value of property, the form of damages awards may not be such as to provide a *continuing* incentive to the polluter to abate. Common law damages can be given only for past losses, although fresh claims can be made by a plaintiff if new infringements occur. Continuing infringements require, in theory, continuing compensation, but the courts are not likely to welcome repetitive claims in connection with the repeated interference. If the court resorts to an injunction then efficiency consequences may follow, as we shall see. Equitable damages, on the other hand, may be awarded for anticipated future interference, but the courts do not allow repeat applications. This implies that, in order to calculate the damages correctly, the duration of the interference must be accurately predicted. Yet once the award has been made the polluter has little incentive to minimize the duration.

Despite these reservations on the form of the damages awards, for the nuisance conflicts between neighbours the practice of the courts may reasonably approximate the compensation flows that would follow from idealized bargains based on polluter liability.[4] But the courts do not restrict their judgements to the award of damages; they often award an injunction to the plaintiff. If an injunction were fully negotiable (in the sense that the polluter could compensate the pollutee to obtain a release) and if bargains were feasible, injunctions would provide no obstacle to socially efficient pollution levels. But if injunctions are imposed in the context of no feasible bargaining, efficiency effects may result. Let us consider these effects in theory to begin with, and then look at the practice of the courts in awarding injunctions.

If the socially efficient level of pollution were zero, that is if marginal pollution cost exceeded marginal abatement cost for *all* units of pollution, a total injunction banning the interference would be socially efficient. But in the more prevalent case, where *some* abatement is socially efficient, but not abatement down to zero pollution, a total injunction is not socially efficient when it is effectively non-negotiable. How do the courts respond to this potentially inefficient restriction of polluting activities? From the layman's point of view *partial* injunctions, which restricted the pollution level but did not eliminate the interference entirely, might seem to offer a compromise. They could be combined with damages awards to compensate victims for any remaining (socially efficient) interference. But in fact partial injunctions rarely seem to be used. The courts do have the power to award damages in lieu of injunction, but their reluctance to 'balance the equities', and compel the plaintiff to accept compensation in place of physical protection, reflects the private law's preoccupation with individuals' rights.[5] Thus Lord Upjohn could argue in the House of Lords that 'an argument on behalf of the tortfeasor [polluter] . . . that this will be very costly to him . . . receives scant, if any, respect' in the process of deciding in favour of an injunction.[6]

Quite apart from their unwillingness to withold an injunction on grounds of social efficiency, there is considerable doubt as to whether, as they are currently organized, the courts are competent to *identify* the socially efficient pollution level anyway. Technical advice from engineers, ecologists or even economists is not called for by the courts (although plaintiffs may of course incorporate the arguments of experts in their case). It is not surprising therefore to find cases in which, for example, a very blinkered view has been taken of the available abatement methods. Here are two examples.

(1) *Sturges v. Bridgman (1879)*

The plaintiff, who was a doctor, purchased a premises whose garden adjoined the property of a confectioner in Wigmore Street who had used noisy machinery for over twenty years. After eight years the doctor built a consulting room at the bottom of his garden, and it proved to be the case that his examinations of patients were disturbed by the thudding of the mechanical mortar and pestles next door. The court found for the plaintiff and granted an injunction.

Even presuming that the marginal damage costs exceeded marginal abatement costs, this decision could have been socially efficient only if the confectioner was able to abate more cheaply than the doctor could (by moving his consulting room). The court did not apparently consider the relative costs of the abatement of the nuisance by either party. Nor did it consider the possibility of a partial injunction, say limiting the interference to certain times of the day to allow uninterrupted examinations part of the time.[7] It is not clear, therefore, that the imposition of the total injunction was socially efficient.

(2) *Bryant v. Lefever (1878)*

The plaintiff was able to light fires in his house without trouble until his neighbours built a wall up past his chimney. The result was that the plaintiff's smoke refused to become an external pollutant and remained in the plaintiff's own house. The court did not applaud this form of pollution control and awarded the plaintiff £40 damages. The appeal judges contemplated the problem of identifying who caused the problem (if the defendant hadn't built his wall . . . , if the plaintiff didn't light fires . . .[8]) and reversed the decision. Was the outcome socially efficient? The court made no apparent effort to identify the socially efficient outcome by obtaining estimates of the cost of abating in different ways (lower wall, higher chimney, etc.). The outcome could have been socially efficient only if the smoke-maker happened to be able to abate more cheaply than the wall-builder and if *no* smoky fires yielded net benefits to society.

What can be concluded from this brief look at the practice of the British courts, the general tenor of which seems much the same today? Damages awards potentially offer the advantages over injunctions of allowing the polluter to respond with his choice of pollution level (thereby averting the worst consequences of penal court injunctions) and of compensating pollutees for all of the pollution cost they incur. Legal practice may fall short of this potential, but private law-suits still offer a useful compensating mechanism within a largely statutory approach to pollution control. Perhaps the major obstacle to the effectiveness of private law in this respect is the heavy litigation costs in cases involving large numbers of pollution victims. There has been considerable discussion of the 'class actions' allowable under the United States Federal Rules of legal procedure as a means of reducing these costs. In a class action a single member of the victim group may file a pollution action, and the case may be joined by other members of the group. But it is optimistic to think that actions of this kind could be introduced in Britain as a major means of pollution control or even of providing full compensation to victims for damage due to pollutants with wide-ranging effects. A recent class action in Los Angeles illustrates the problem.[9] The action was filed on behalf of the inhabitants of Los Angeles County against 291 corporations allegedly causing air pollution. The court dismissed the action because, among other reasons, if did not feel capable of 'balancing the interests of the inhabitants of the Los Angeles Basin against the needs of productive industry in the same area'; in other words, it could not evaluate the social efficiency of a complex set of air pollutants emanating from many sources. This suggests that the court was thinking in terms of the problem of controlling pollution rather than of compensating victims. For an award of damages in lieu to have been made only (!) the pollution costs would have had to be estimated; but then for the individual polluters' shares of the damages payment to be determined a share of the pollution 'pool' would have had to be attributed to each of them. Clearly, the more pervasive a pollutant the harder the task of achieving even some compensation of victims through court awards. But it remains important that lawyers explore other means of providing compensation in court. Perhaps some form of test case, followed by summary treatment of damages claims by other victims, offers the best prospect for the British legal system. For reliance on statutory intervention for the bulk of pollution control leaves unresolved the problem of uncompensated harm resulting both from the pollution that does not infringe the statutory standards and from the pollution that is in violation of these standards. Such compensation ideally would be the meat of private law-suits in the context of an interventionist, centralized pollution control policy.

5.3 STATUTE LAW: LEGISLATION AND ENFORCEMENT

Centralized pollution control policy in Britain and the United States has been predominantly, but not exclusively, in the form of regulated standards of one kind or another.[10] In Britain the initial impetus to statutory intervention came from the nineteenth-century movement to protect public health in a newly industrialized economy. The instrument of control was the prohibition of 'nuisances' which were 'prejudicial to public health'.[11] This rather general attack on pollution has subsequently largely been replaced by more specific controls, but it remains relevant to pollutants that are difficult to quantify, in particular unpleasant smells and noise that is only imperfectly measured by decibel ratings.

The attack on pollution has been sharpened by the gradual shift from statutory nuisance to the legislation of emission standards. This shift required the legal acceptance of the principle that an activity could be restricted if it was *presumed*, as distinct from proved, to be harmful. This acceptance was forthcoming as a series of reports established the link between the emission of pollutants and untreated sewage and the health of those affected.[12] Several different methods of establishing the standards have been tried in Britain.

(1) *Nominal prohibition with weak enforcement*

The 1876 Rivers Pollution Act introduced the practice of prohibiting *all* forms of river pollution while simultaneously severely limiting the circumstances in which the law could be enforced. In particular, enforcement was permitted only in cases where there would result no material injury to the interests of the polluting industry! Strictly interpreted, this implies that enforcement requires abatement costs to be negligible. The economist's objection would clearly be that this rules out the balancing of interests, of abatement costs against pollution costs, which social efficiency requires. A lawyer, on the other hand, might be concerned particularly with the two characteristics of the Act that it embodied Parliament's attempt to obstruct the application of the law it was creating, and that polluters could not know in advance (in the absence of statements by the enforcers) what would constitute an infringement of the law as made effective by the enforcement agency.

(2)   *Legislated uniform standards*

As a means of reducing the uncertainty over the level at which standards will effectively be established, the legislature may itself specify the standards. This approach is increasingly common practice in the United States and other European countries,[13] but in Britain its adoption had largely been restricted to air pollution.[14] The 1912 Royal Commission on Sewage Disposal felt that standards were not feasible for many pollutants whose effects were not established, but it did recommend a guideline of 3 parts suspended matter per 100,000 for sewage effluent, for sanitary (later river and water) authorities to follow. The 1949 Report of the River Pollution Prevention Sub-Committee of the Central Water Advisory Committee took the view that local variations made local standards preferable to national standards, and it recommended a system of legislated local standards. The 1951 Rivers (Prevention of Pollution) Act gave river boards the power to establish local standards on any river through bye-laws, but no such laws were ever confirmed owing to the difficulty of formulating standards for all pollutants. This experience led to a lack of enthusiasm for the rigid application of legislated standards on the part of legislators and administrators, and to the adoption of a more flexible 'consents' method of control for water pollution (see p.160 below), which some people regard as characteristic of the British approach to pollution control and which has been endorsed by the Royal Commission on Environmental Pollution.[15]

(3)   *Individual standards*

In the control of both air and water pollution increasingly the British pollution control agencies have tended to opt for flexibility in the sense of retaining discretion over the severity of the standard to be applied to particular polluters. In the case of air pollution, Parliament delegates to the Alkali Inspectorate and other bodies the power to establish individual standards for polluters that satisfy the general requirement that emission levels do not exceed those attainable with the 'best practicable means' in terms of technological and managerial feasibility, but it does not require that any economic efficiency criterion be applied. There is no reason to think, therefore, that the 'best practicable means' criterion offers a way to allow a flexibility in standards based on marginal abatement cost and marginal damage cost differences between polluters. The criterion *could* be interpreted in this way, but it is well known that the Alkali Inspectorate lacks the expertise to undertake the requisite cost studies to relate the standard set to the best estimate of the individual polluter's socially efficient level of emission.[16] There are no economists employed in the Inspectorate! The Royal Commission is correct to suggest that 'at its best [best practicable means] connotes a rigorous analysis of the objectives and consequences of air pollution control. At its worst the term can be used as a catchword to conceal the absence of any such analysis'.[17] More worrying, the best practicable means criterion can be used by the control agency as an excuse for inactivity. Unless an explicit cost–benefit analysis of a local factory's emissions is undertaken and published, how are those affected by the pollution to distinguish between socially efficient and socially *in*efficient pollution? How are they to check the usual proposition by the local Alkali Inspector that 'best practicable means' are being employed? And how are they to know whether 'practicable' is being interpreted as 'efficient in view of the benefits and costs to society' or as 'financially feasible given that the firm's managing director is reluctant to postpone his expensive holiday and his wife's new Rolls Royce'? The best practicable means has potential for an efficiency analysis of pollution abatement, but the traditional control agencies in Britain may well be unable to provide it unless some radical changes in their structure and expertise are undertaken.

In the case of the control of river pollution, the enforcement agencies (the regional water authorities) have the power to set the individual standard which a polluter must meet. The polluter must obtain a 'consent' to pollute, and the consent that he receives specifies the quantity and concentration of the effluent he may emit. It is a criminal offence to fail to meet the conditions laid down in the consent or to pollute without a consent. The consent instrument incorporates the flexibility of relating the standard to local conditions, such as the expected effect of an emission on the downstream river quality. An individual standard is negotiated with the polluter but once this has been done the discharger faces a formally stated requirement. In this respect the consent method differs sharply from the nebulous requirement that best practicable means be used to avoid pollution. Changes in the terms of any particular consent can be made only through formal proceedings; it cannot be done through a friendly chat with the local inspector. The discretion exercised by the regional water authorities is clearly substantial. Not only do they decide who to prosecute for failure to comply with consent conditions, but in setting those conditions they are effectively defining what is a legally permitted activity.[18] A beginning has been made to increasing the public accountability of the local enforcement agency by the requirement in the 1974 Control of Pollution Act that some types of consents be registered. This may provide a means of reducing the regional variations in the stringency of consent conditions which do not reflect differences in the abatement and damage cost situations of polluters, but which reflect rather differences in the diligence of the various water authorities.

From this outline of the legal structure for air and

water pollution control in Britain it is apparent that it is not current British policy to press for national uniform standards. A very great deal of flexibility is built into the system, the disadvantage of which is that it can conceal inactivity on the part of the law-enforcers who have a substantial degree of discretion in standard setting and enforcement. Certainly neither air nor water pollution control agencies have been keen even to prosecute the violators of standards.[19] Krier argues in relation to the United States that a system that allows regional variations in pollution control may give too much leverage to polluting interests.[20] He argues in favour of centrally imposed uniform standards which allow regional pollution control agencies to set more or less stringent standards if they can demonstrate lower- or higher-than-average abatement costs. This, as he says, would alter the balance of power between polluting and anti-polluting interests. This is because the polluting interests would then have to overcome the inertia of the legislative process (bear the burden of proof) in order to bring about a local slackening of standards. Under the ultra-flexible British system, the inertia has to be overcome by those who favour a tightening of standards generally, in the absence of any obligation of most regional agencies to exert control to a centrally specified level.

The second problem with the British approach in practice is that there is little evidence that the air and water pollution enforcement agencies have the inclination, or the expertise, to orientate the variations in the individual standards that they set towards the minimization of the sum of abatement costs and pollution costs, or even towards the minimization of total abatement costs for polluters in similar environments. There has been to date virtually no attempt to estimate and publish cost-of-abatement figures for particular pollutants in particular localities. There *may* be a wealth of such cost information inside the Alkali Inspectorate, the Water Authorities and other agencies, but it is more likely that standards are set with no firm cost information to hand.[21] In the absence of such information the enforcement agencies are naturally in a weak position to respond to the normal response of polluters when faced with the prospect of tightening standards, namely that the costs are prohibitive and will lead to unemployment and so on.

In contrast to the British approach the United States is pressing ahead with a harder-hitting, nationally more uniform attack on pollution which is more transparently *active* but which has problems of its own. In the case of air pollution, the uniformity relates to the air quality target (ambient standard) set rather than to the emission standards imposed on particular polluters. A uniform ambient standard implies the strictest emission standards for those polluters whose pollutants most seriously affect air quality, but the implied emissions standards are not sensitive to differences in polluters' abatement costs.

The legislative developments in the United States have been clearly outlined by Kneese and Shultze, so they will not be described in detail here.[22] We limit attention to one or two characteristics of the current centralized pollution control machinery.

In the first place, the emphasis of much of the United States legislation has been on *ambient* standards, and this has given rise to problems of proof that are not common with British legislation, which has to a greater extent relied on emissions standards. The difficulty with an ambient standard is that with multiple pollution sources it may be hard to establish legal proof of the connection between particular polluters' emissions and the quality of air or water in the locality. Not surprisingly, the United States legislation of the 1970s has shifted the emphasis towards emissions controls. For example the 1970 Clean Air Amendments, while retaining overall ambient air quality targets, gave the Environmental Protection Agency (EPA) power to set emission standards for hazardous pollutants. For many pollutants from new industrial sources, the EPA is to require the 'best adequately demonstrated control technology' to be used, which is reminiscent of the British 'best practicable means', which we have seen to be problematic in practice. In addition the new legislation established emissions limits for automobiles, which so far the British have made no attempt to do except on new vehicles.[23] Interestingly, the Royal Commission on Environmental Pollution does not even list vehicle emissions under 'matters requiring attention', although it does say that 'we may expect' limitations on the use of cars in urban areas.[24]

In water pollution control also the United States 1972 Water Pollution Control Act reflects a move away from the ambient standards of the 1965 Water Quality Act towards emission standards. Taken with the 1970 Clean Air Amendments, this new legislation represents an ambitious attempt to reduce air and water pollution nationwide, but Kneese and Shultze adopt a very antagonistic attitude towards it.[25]

In considering their views we should note another major difference between the British and United States approaches. In the United States most of the enforcement agencies have been amalgamated into the EPA, whereas in Britain enforcement duties remain fragmented between many bodies. What is more, the EPA is subject to tighter legislatively specified duties than are the British agencies, who retain greater discretion over standard setting. One's instinctive reaction therefore is to feel that any indictment of the likely enforcement performance for the United States legislation will apply *a fortiori* to the British case. Kneese and Shultze's scepticism about the legislation appears to be based on two main arguments. First, the EPA is charged with establishing national emissions standards that are technologically feasible, and that do not impose 'unreasonable' abatement costs. It is hardly surprising to find that EPA's interpretation of 'feasible' and 'reasonable' do not go unchallenged by large polluters, and yet the fact that challenges do take place is regarded by Kneese and Shultze as a reason for profound scepticism. Inevitably court cases have been arising over the interpretation: how could it be otherwise? Any system of control that imposes costs on the polluter will make some polluters regard litigation as worthwhile. Pointing to some cases of litigation is not in itself sufficient grounds even for moderate scepticism; it certainly is not the firm basis for total disillusionment with the system that Kneese and Shultze seem to think it.[26] The fact that some people accused of murder may prosecute for wrongful arrest is not sufficient grounds for thinking enforcement of the law against murder will be negligible. The fact that some people challenge their tax assessment does not make us abandon income taxes. After all, the court cases only reflect the attitude of those who are disinclined to comply with the regulations. We need evidence also on the polluters who *do* comply, without litigation.

This opposition to the ultra-sceptical view of the legislation is not intended as a whitewash. The enforcement of any legislation is problematic, and the enforcement of legislation that restricts activities that have not customarily been seriously restricted is more problematic than most. But a balanced view of the legislation's prospects is not aided by carefully selecting the most pessimistic pieces of evidence available.

The second basis for scepticism about the legislation that is discernible in Kneese and Shultze's argument is the suggestion that the costs of abatement implied by the Acts are very large.[27] They calculate the costs as well as they can, given the evidence available (which is infinitely better than the evidence currently available on the costs of compliance with British legislation). Their conclusion is that the total abatement costs might amount to 10 per cent of the growth in *per capita* national income (for a time period whose duration is unclear). As they correctly point out, this places in perspective the arguments that have been made for zero growth as a means of controlling pollution. They also plausibly argue that the costs can be kept down by not pressing for compliance too quickly. But why should the minimum level of costs at which compliance might be achieved be described as 'massive', and be regarded as another reason for scepticism? It is true that the higher the costs the more likely it is that enforcement will meet opposition. No doubt generous standards that polluters can comply with at negligible cost would be easier to enforce. Kneese and Shultze do not suggest that the abatement costs relate to target pollution levels that are too *low* for social efficiency. The real reason the costs are enumerated, one suspects, is to provide them with a starting point for the advocacy of charges. They re-iterate the common argument that studies of certain river basins indicate that charges would reduce the total abatement costs associated with particular standards. The extent of the reduction is not indicated. The application of the argument to air pollution, when the quoted evidence is for water pollution, is not provided with any justification. But most curious of all, when charges are discussed the enforcement problems seem to evaporate, despite the fact that those charges would be in support of standards as hard-hitting as those in the recent legislation.[28] It may of course be that a case for charges can be made in particular contexts, where abatement costs differ widely for particular pollutants, as we saw in chapter 4. But the *general* advocacy of charges as an 'alternative strategy' scarcely stands up on the basis of the assertions used in this critique of the current legislative position in the United States.

Finally, the ultimate effectiveness of the approach in the United States remains to be seen. At least the system there is open, in the sense of having some explicit declared abatement targets. This contrasts with the covert administrative discretion of the British system. Perhaps most important in the long run, the United States approach is generating information on the likely costs of alternative abatement strategies. The current position in Britain is that, in the absence of *explicit* targets, little attempt is being made by the Department of the Environment to induce the detailed cost studies which will facilitate the future tightening of pollution standards in Britain.

Notes

CHAPTER 1

1   For example Meadows *et al.* (1972), but for a critical appraisal see Cole *et al.* (1973).
2   On discernible pollution trends see the Royal Commission on Environmental Pollution (1974), and Lee and Saunders (1972). For information on quantities of pollutants emitted in the United States and other countries see Kneese (1977), chapters 2 and 3.
3   See, for example, Lipsey (1975), part 7.
4   For analyses that treat assimilative capacity more formally see Pearce (1976b), chapters 3 and 4, and Ayres and Kneese (1969).
5   See the introductory books by Edel (1973), Barkley and Seckler (1972), and Seneca and Taussig (1974).
6   For example, Baumol and Oates (1975), and Maler (1974).

CHAPTER 2

1   Surveys can be found in Mishan (1971) and Dick (1974).
2   See section 1.2.
3   See section 2.4. on property rights.
4   It seems inconsistent to maintain, as Baumol and Oates do (1975, pp. 17–18), that external costs must lie outside the polluter's decision calculus *and* that the absence of a flow of payments is not a necessary condition for the existence of an external cost.
5   These are often referred to as *pecuniary* external costs to distinguish them from the direct imposition of utility losses or production cost increases, often called *technological* external costs. The exclusion of pecuniary external costs means that in this book the term 'external costs' will refer specifically to technological external costs.
6   That is to say there is nothing inherently different in an engineering sense.
7   For example Mishan (1971); Dick (1974).

168 *Notes*

8 See sections 2.4 and 3.1.
9 It might be possible to reduce the concentration of the effluent by smoothing a given output through time to avoid output peaks, but this also involves costs such as stockholding. For simplicity the output level will be assumed to determine the effluent level from a given production process.
10 This is generally true whether or not the market is competitive as is assumed in figure 2.1.
11 Some textbooks (e.g. Dick, 1974) proliferate models of polluter behaviour by distinguishing different types of polluters (and pollutees).
12 To keep the problem two-dimensional these other inputs will be limited to labour, but 'labour' can be thought of as an amalgam of non-environmental inputs.
13 The reason why the isoquants Q_1, Q_2, Q_3 are not smooth and convex to the origin as in conventional neoclassical (textbook!) analysis is that we have displayed a small number of processes. This is a simple example of a linear programming formulation of production. For a very clear explanation of the basics of linear programming see Baumol (1972a).
14 See Pearce (1977b, chapter 8; 1976a).
15 In other words at A th marginal product of the environmental input is zero, as shown by the isoquants becoming flat (see Burrows, 1977). The curved isoquants in figure 2.3 indicate that we are now assuming many intermediate processes between, $P1$, $P10$ etc. to exist. We shall also assume the production function to be linear homogeneous.
16 Note that the cross-over of MAC output cut (2) and MAC process switch at E_8 is irrelevant to the polluter's choice of abatement method. This is because the MAC output cut curve reaches position (2) only *after* the process-switching which has abated from E_{10} to below E_8.
17 Lancaster (1966a, 1969).
18 The most thorough exploration of these aspects is in Lancaster (1966a, b).
19 In the case of external costs it need not, of course, be only the utility functions of the consumers.
20 For simplicity it is assumed that, in Lancaster's terminology, there is a one-for-one relationship between goods and consumption activities.
21 We are obviously not regarding as 'average' those with conservationist fervour, especially those whose altruistic enthusiasm takes the form of wishing to *pay* (as distinct from campaign) for the protection of other pollutees.
22 One result of this conceptual similarity is that mathematical models of external costs treat the impact of pollution on consumption and production symmetrically. Thus the utility functions of the

i consumers are often written as

$$U_i = U_i(x_{1i} \ldots x_{ji} \ldots x_{ni}, s_i), \quad \frac{\partial U}{\partial x_{ji}} > 0, \quad \frac{\partial U}{\partial s_i} < 0$$

where $x_1 \ldots x_n$ are goods, and s is the amount of pollution 'received' by a typical consumer. Similarly the production functions of the k producers affected by pollutants are written as

$$Q_k = k(y_{1k} \ldots y_{jk} \ldots y_{nk}, s_k), \quad \frac{\partial Q}{\partial y_{jk}} > 0, \quad \frac{\partial Q}{\partial s_k} < 0$$

where $y_1 \ldots y_n$ are factors of production, and s is the amount of pollution 'received' by a typical producer.

23 Remembering that the 'amount' is a composite index of quantity and concentration.

24 Technically stated, the total cost function is twice differentiable and convex.

25 See Baumol and Oates (1975), chapter 8); Portes (1970). Starrett and Zeckhauser (1974), Slater (1975) and Gould (1977). As usual the literature contains a degree of obfuscation. Thus for 'detrimental externalities and non-convexities in the production set' (Baumol and Oates, 1975, chapter 8) the uninitiated should read 'severe pollution costs and the downward-sloping marginal pollution cost curve'.

26 Baumol and Oates (1975), Portes (1970) and Starrett and Zeckhauser (1974) challenge assumption (1); Slater (1975) disputes assumption (2).

27 The total cost curve would have greater curvature if in addition the value of a marginal unit of damage increased as total damage increases.

28 Depending on the curvature of the total pollution cost curve segments to the left and right of E_1, the corresponding marginal pollution cost curve segments could be as shown in (f), or both linear, or both inverted (yielding a smooth MC curve with a maximum at E_1).

29 Slater (1975), p. 868.

30 See Baumol (1972a, chapter 12, section 11).

31 Odd-shaped functions can produce indeterminate results as the reader can verify if he follows through the analysis of section 3.1 using flat marginal abatement and pollution cost curves, which either coincide or fail to intersect.

32 Kapp (1970; p. 85 in Wolozin, 1974).

33 The existence of families of curves is of course common in economics because our diagrams try to represent in two dimensions relationships containing more than one independent (explanatory) variable.

34 cf. Kapp (1970; p. 86 in Wolozin, 1974). Kapp also criticizes economists' concentration on the analysis of markets, a point that

we are inclined to agree applies to *some* economists. However, the theory of external costs is not limited to the analysis of markets; as we shall see in the next chapter, it provides a useful basis for the evaluation of various forms of government intervention.

35 This has not deterred some economists from devising sophisticated names for marginal effects (e.g. Pareto-relevant or non-separable) and non-marginal effects (non-Pareto-relevant or separable) and spelling out the differences in the consequences of the two types of effect. See Davis and Whinston (1962) and Wellisz (1964).

36 See Davis and Whinston (1962) and Wellisz (1964) for an analysis of reciprocal externalities.

37 Students unfamiliar with these terms will find the analysis here rather terse and are advised to read Millward (1971, pp. 131–41) and then refer to the fuller discussions in Samuelson (1954) and Head (1962).

38 Head (1962 section 4).

39 Baumol and Oates (1975, p. 46, n. 22) agree that cases of private bad (their 'depletable externalities') are hard to come by.

40 Remembering that if B indulges in pollutee abatement, say the installation of soundproofing in his house, then the cost of abating is still a pollution cost to him and presumably a cost that is equally available to A if he prefers it to the noise.

41 A lucid and more detailed discussion can be found in Dales (1968, chapter V).

42 The polluter pollutes if and only if for some units of pollution

$$p(P + f) < a$$

where p is the probability of paying, P the pollution cost, f is the fine and a the abatement cost.

43 For some references to legal studies see chapter 5.

44 Even though private land-owners may retain *some* rights in clean air (e.g. protection from the neighbour's unreasonable bonfire nuisance), the limitless air above private property is essentially a common resource; and the back garden oil prospector will soon find that his rights to the riches of the soil are extremely limited.

CHAPTER 3

1 The reader might test this statement by trying to find any substantial discussion of justice in the available textbooks by economists (e.g. Baumol and Oates, 1975; Dick, 1974; Pearce, 1976b; Seneca and Taussig, 1974; Victor, 1972 or in the writings of lawyers (e.g. Posner, 1972, and Calabresi and Melamed, 1972). It seems that the lawyers are suffering from tunnel vision as a result of discovering the delights of the economists' 'rigorous' theory of efficiency. It

seems strange to hear lawyers argue, for example, that little can be said of the consequences of pollution control for justice unless we indulge in 'transcendental meditation' (Calabresi and Melamed, 1972, p. 1102).

2   Readers not familiar with basic efficiency analysis should consult Millward (1971, chapter 2 and section 4.1).

3   Most textbook analyses are of this kind and are inclined to tempt the authors into the advocacy of policies purely on the basis of the net gains yielded. Baumol and Oates (1975, chapter 3, section 4 and chapter 4), for example, argue that compensating pollutees may be inefficient and proceed to advocate a tax-without-subsidy solution with no consideration of justice whatsoever. More of this in section 3.4.

4   Rawls (1971, chapter II) and Fletcher (1972).

5   There are differences of detail between the various cases which we shall not pursue. Repetition of the basic model is inclined to be tedious, but for those who seek a more complete exposition a convenient survey can be found in Dick (1974, chapters 3 and 4). Note that the type of external cost that Dick refers to as 'nuisance' is our case of fixed technology for the polluter; his 'pollution' or 'congestion' type is our case of flexible technology. His use of the terms 'nuisance', 'pollution' and 'congestion' is misleading because it differs from common usage: for example, 'pollution' does not cease to be pollution because the polluter's technology is rigid. Also, Dick's limitation of the term 'abatement' to abatement through process switching goes against normal usage.

6   See p. 35 *et seq.*

7   We could also consider whether the analysis can explicitly incorporate adjustments in the *pollutee's* process or location to reduce the damage resulting from a given amount of pollution. However, to simplify the analysis the cost-minimizing reaction to a given level of pollutant is left implicit; in other words, the marginal external cost curve is drawn on the assumption that pollution costs for a given amount of pollution are minimized.

8   Formally the total cost functions of A and B are

$$C^A = C^A(q^A), \frac{dC^A}{dq^A} > 0 \; \frac{d^2C^A}{d(q^A)^2} > 0$$

$$C^B = C^B(q^A, q^B), \frac{\partial C^B}{\partial q^B}, \frac{\partial C^B}{\partial q^A} > 0; \frac{\partial^2 C^B}{\partial (q^B)^2}, \frac{\partial^2 C^B}{\partial (q^A)^2} > 0.$$

Models based on functions of this kind are common in the literature. See for example Marchand and Russell (1973), Gifford and Stone (1973) and Browning (1977).

9   If $Opab < O'c'd'$ and q_1^A were the smallest non-zero output that A could produce, then *no* output by firm A yields net gains to society. This is known as the corner solution q_0^A, q_0^B.

10   If for every level of A's output up to q_n^A the marginal benefit
     (profit) to A from an increase in q^A exceeded the marginal loss
     (of profit) to B, then the socially efficient situation would be for
     A to produce (and pollute) to the level that maximizes its own
     profit, namely q_n^A. The outcome is another extreme case, q_n^A, q_n^B
     being socially efficient.

11   Formally, the first-order conditions for the maximization of A and
     B's joint profits, $\pi^A + \pi^B = P^A q^A - C^A(q^A) + P^B q^B - C^B(q^A, q^B)$, are

$$P^A = \frac{dC^A}{dq^A}(q_2^A) + \frac{\partial C^B}{\partial q^A}(q_2^A, q_2^B)$$

$$P^B = \frac{\partial C^B}{\partial q^B}(q_2^A, q_2^B).$$

     For both firms price equals marginal social cost, but in B's case
     marginal social cost and marginal private cost are equal. See, for
     example, Gifford and Stone (1973) for further discussion of these
     conditions. The conditions derive from the solution of the simultan-
     eous model represented by A's and B's profit equations. Clearly
     we can discuss A's behaviour only given B's and vice versa; in
     the diagrammatic explanations we describe the behaviour of the
     polluter which is socially efficient if the *pollutee* reacts in a socially
     efficient manner also.

12   But see n. 15 below.

13   The activities are reciprocal in the sense that from the viewpoint
     of society as a whole a change in the level of one activity alters
     the profitability of the other. This does not imply that A pollutes
     B's activity *and* B pollutes A's. We have limited attention to uni-
     lateral pollution.

14   The loss of profit to B can be the result of pollution damage
     or of the increase in the firm's costs due to expenditure on damage-re-
     duction.

15   The corner solution q_0^A, q_0^B (see n. 9 above) results if $MPC^B > MAC^A$
     at all levels of q^A and q^B. The other polar case q_n^A, q_n^B requires
     $MPC^B < MAC^A$ for all levels of q^A up to q_n^A (see figure). These
     two cases indicate that $P = MSC$ is not *necessary* for social efficiency,
     although it is *sufficient* in the simple model with the convex
     joint profit curve as shown in figure 3.2 (b).

16   In the private bad case, therefore, compensation paid to the pollutee
     does improve efficiency. With private bads two pollutees, B1 and
     B2, are not obliged to receive the same quantity of pollutant each,
     and an increase in B1's receipt of pollution reduces the amount
     available to B2 (and B3, B4, etc.). Setting a negative price (compensa-
     tion) on the pollution leads to an efficient allocation of the pollutant
     between pollutees. On the formal conditions for efficiency in this
     case see Baumol and Oates (1975, pp. 45–8).

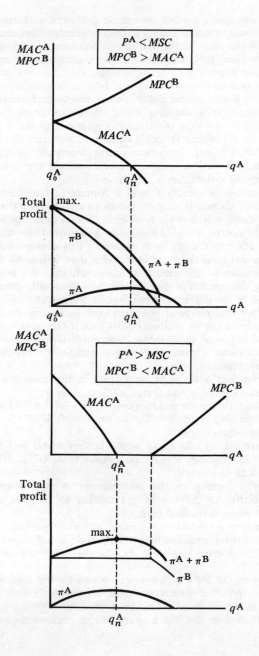

17 It will also provide other (potential) pollutees with the incentive to keep away from the locality unless the benefits from their activity outweigh the costs to the polluter of adjusting to the higher level of internalized pollution cost that their entry to the area would entail. See Baumol (1972b), section III).

18 We say not *necessarily* because if firm B is indifferent between q_0^B and q_2^B it may choose either, or any output in between.

19 See Pearce (1967b, chapters 2 and 4), who extends the analysis to discuss the problem of accumulating non-degradable pollutants (called stock pollutants) such as cadmium, mercury and lead.

20 B's socially efficient output is of course dependent on A's, so that, as in the conventional model, we are drawing the solution to a simultaneous system. See n. 11 above.

21 These conditions identify a local maximum (a hump on the total joint profit curve). It may be (as in case 2 below) that there is another point which yields even higher social gains, in which case q_1^A would not be a *global* maximum and would not therefore be socially efficient. Clearly, with models of this degree of complexity the sufficient conditions are little more than a lengthy list of the characteristics of the socially efficient solution. We are close to reaching the conclusion that the outcome is socially efficient when it has all of the characteristics of social efficiency!

22 Note that with marginal pollution cost increasing with the level of pollution a corner solution occurs only if the MAC^A and MPC^B curves don't intersect; with the downward-sloping marginal pollution cost curve a corner solution is compatible the existence of an intersection.

23 This point has not always been clear in the literature (compare for example Baumol and Oates, 1975, chapter 8).

24 In quadrant (b) the marginal tax cost (MTC) curve would be horizontal and passing through *l*. Up to output q_1^A $MAC^A > MTC$ and polluting is worthwhile.

25 The marginal tax cost curve would be horizontal and starting at point *i'* in quadrant (e). At no output is $MAC^A > MTC$ so no pollution is worthwhile.

26 The polluter group in any one industry is still assumed to be small relative to the size of the market so that the horizontal demand curve is retained here.

   Some care is needed in summing individual polluters' marginal abatement cost curves and individual pollutees' pollution cost curves but we shall leave this aside for the moment: see section 4.2.2 below.

27 Once again the MPC (q_j^B) curve is drawn on the assumption that firm B responds efficiently to interference from A and adjusts its output to the socially efficient level q_j^B. For simplicity we assume that firm B does not find a change in *its* technology worthwhile.

28 The marginal pollution cost curve under fixed technology, $MPC^B(q_2^B)$ is lower than that under flexible technology, $MPC^B(q_j^B)$, because in the latter case B's socially efficient output is higher, and the damage for any level of E^A is therefore higher. For the same reason $MSC^A(q_2^B)$ is lower than $MSC_1^A(q_j^B)$ in quadrant (a).

29 See figure 2.3(a) above.

30 This curve is the same as in figure 2.1, quadrant (b) with the axes reversed.

31 There may be several processes and locations available, but we shall consider only one process in different locations.

32 For any location the *marginal* cost of moving is the cost over and above the cost of moving to the next most polluting location. If there exists a location X which is less polluting and less costly to move to than another, Z, then X dominates Z and Z is eliminated from the set of locations from which a choice is made.

33 The evaluation of abatement methods is complicated in practice by the fact that both polluter *and* pollutee abatement in each of these forms should be compared.

34 The subsequent discussion will retain the assumption of competitive factor markets and concentrate on imperfections in goods markets.

35 Buchanan (1969).

36 Baumol (1972b, p. 308).

37 The following outline and evaluation of Buchanan's argument is a summary of Burrows (1978a, section 2).

38 Note that the capital Qs indicate that we are now talking about industry rather than firm output and pollution levels.

39 The $MSC^A(q_2^B)$ curve is drawn on the assumption that the pollutee(s) responds efficiently to the monopolist's pollution. Note that q_2^B is higher than q_j^B in the competitive case because the pollution level is lower under monopoly.

40 The socially efficient output under competition is found from the intersection of the demand curve and a lower marginal social cost curve $MSC^A(q_k^B)$, where $q_k^B < q_2^B$ because the pollutee reacts to the higher competitive output. Buchanan's argument relies on the assumption that this competitive output does in fact exceed the monopoly output $Q_n^A(m)$. See Buchanan (1969) and Burrows (1978a, fn. 12).

41 If marginal pollution cost were very high so that the $MSC_1^A(q_2^B)$ curve intersected the demand curve to the left of point d, a net social loss would not result. If monopoly power were weak then individual polluters' demand curves would be relatively flat so that before pollution control the industry output would be near the competitive level $Q_n^A(c)$, and abatement would be likely to involve a move *towards* the socially efficient output.

42 We shall ignore the second order welfare effect of an output cut which results from the increase in cost due to a process switch.

43 Our discussion of the nature of property rights has benefitted from the beautifully lucid analysis in Browning (1977). He correctly chastizes many economists for lack of care in the definition of the damages incurred by pollutees.

44 We have throughout our efficiency analysis assumed that there is no benefit to other pollutees from a reduction in the level of pollution suffered by B. That is, pollution for us is generally a public bad. If this were not so compensation payments that related to the level of B's activity could be socially efficient. This is the case of depletable external costs that we may regard as rare if not mythological. On the analysis of such cases see Baumol and Oates (1975, chapter 4, for example p. 53).

43 It is assumed here that there is no threat-making by A to extract extra compensation; but see p. 89.

46 Originating in Coase (1960). Marchand and Russell (1973) have argued that with costless bargaining the allocative outcome will not be symmetrical with respect to rights. However, their results hinge on the open-ended form of liability, so that they do not challenge the Coase theorem if polluter liability is based on socially efficient pollutee behaviour.

47 The right-hand shaded area in figure 3.2 is the net social *loss* from output q_2^A to q_n^A or the net social *benefit* from abatement from q_n^A to q_2^A. The left-hand shaded area similarly can be viewed as the net social loss from abatement from q_2^A to q_0^A, or the net social benefit from A's activity up to q_2^A.

48 Where the polluter and pollutee are firms rather than individuals it will be the relative wealth positions of their shareholders and the consumers of their products that will be affected via the impact on profits and prices.

49 There is evidence, for the United States at least, that pollution *is* regressive in its impact: see Freeman (1972).

50 But see the qualification noted on p. 92

51 A fuller discussion of some of the obstacles can be found in Dick (1976). But our interpretation of some of the points differs from his; for example, we would attach less significance to the results of Marchand and Russell than he does.

52 For those who enjoy the intellectual exercise Dick (1976) provides a more general discussion of the dispute over symmetry.

53 See, for example, Demsetz (1971), who postulates that competition between extortionists will drive the 'compensation' demand for fictitious damage to zero. There is nothing logically wrong with statements of this kind; but does it not seem undesirable to use the world's limited intellectual resources to debate such obviously irrelevant theoretical points?

54 The British government has implemented a notification procedure for the dumping of toxic wastes, but 'there still remains considerable uncertainty about the amount of wastes being produced' (and pre-

sumably being largely dumped): Royal Commission on Environmental Pollution (1974), p. 19.
55 See Mishan (1967), who uses this as an allocative argument in favour of imposing polluter liability.
56 See Burrows (1970, p. 45).
57 See pp. 84–5 above.
58 See pp. 35 *et seq.*
59 See p. 89.
60 See, for example Gifford (1977, p. 12).
61 Pp. 75–80 *et seq.*
62 Pp. 78–80.

CHAPTER 4

1 'Not necessarily' because it is possible to devise control policies that shift the burden to others, for example polluter-subsidies: see section 4.2.4.
2 To complicate matters further some people refer to effluent charges as 'the polluter-pays principle': see for example the Minority Report in Royal Commission on Environmental Pollution (1972) by Zuckerman and Beckerman.
3 Again, the literature is plagued with a variety of names for the same policy; for example Dick (1974) refers to 'directives', Rose-Ackerman (1973) to 'allocation regimes' and 'non-market schemes', Rowley (1974) to the 'fiat solution', Baumol and Oates (1975, chapter 10), to 'direct controls'. In the context of British river pollution control the quotas are known as 'consents'.
4 The literature contains many mathematical demonstrations of this result which we shall explain with diagrams; see for example Baumol (1972b, p. 311) and Baumol and Oates (1975), chapter 3, section 3).
5 For simplicity the diagram assumes that one unit of output with a fixed technology generates one unit of pollution, so that the charge is Ot per unit of output as well as Ot per unit of pollution.
6 At E_j^A marginal abatement cost and marginal pollution cost are equal.
7 If the costs of administering a pollution charge were sufficiently in excess of those for an output tax, however, the efficiency-gain advantage of the pollution charge might be outweighed by administration costs. In this case the output would be more socially efficient than the pollution charge. Once administration costs are introduced, however, one must consider whether *any* instrument produces net social gains after the deduction of administration costs. Thus, the net social gain from abatement from q_n^A to q_2^A in figure 3.1 (a) would then be *elm* minus the total cost of employing the policy

instrument used to induce the move to q_2^A. This sum *could* be negative.

8   The succeeding analysis can be modified to allow for the fact that some interference may be accepted as just, so that less than full compensation for interference also could be just. Our assumption in the text has the advantage of emphasizing the *possible* trade-off between efficiency and justice. See chapter 3, p. 52–5.

9   See chapter 3, n. 44 and the reference therein.

10 See chapter 3, pp. 62–3; Browning (1977, pp. 1,286–7), and Holterman (1976, section V).

11 Not all of the practices of British courts are as well conceived as the limitation of liability to interference with the plaintiff's *reasonable* activity: see chapter 5.

12 See chapter 3, pp. 69–70, on the individual polluter's marginal pollution cost curves.

13 Dolbear (1967, section III).

14 The excess revenue could in principle be eliminated by giving polluters exemptions up to a certain pollution level. For example, a charge of *wv* combined with an exemption of OE_e in figure 4.2 yields the revenue *xyvw* which is equal to the total pollution cost *Ovw*.

15 The payment of a lump-sum subsidy to polluters out of the excess revenue would prevent this profit-squeeze effect: Dolbear (1967, p. 99).

16 In addition, if the firm perceived the intra-marginal charge as a component of its marginal cost it would alter its output level. We shall assume here that this is not the case, but see Burrows (1977) for an exploration of the implications of this perception.

17 In Britain the control of land use operates mainly through the granting or refusal of planning permissions. In the United States similar control is exerted through zoning ordinances. In both countries the powers are vested in local government.

18 It is quite easy to allow the discussion of policy instruments to move from the positive analysis of policy impact in qualitative terms to a normative advocacy of particular instruments based on casual empiricism. We shall try to resist this temptation, but see for example Ellickson (1972, section III) for an analysis of the social efficiency of zoning which is permeated with empirical judgements concerning the size of pollution reductions owing to zoning (rather small) and of the abatement costs involved (rather large). No firm empirical evidence is offered to support these judgements, but they are used as a basis for a wholehearted Chicago-style advocacy of market solutions to pollution problems.

19 See Mishan (1968, chapter 8).

20 Ellickson (1972, section IV *et seq*).

21 See Elickson (1972, p. 711).

22 Ellickson (1972, pp. 703–4); Seneca and Taussig (1974, p. 214).

23 The consequences of abatement costs for the general distribution of income may provide another reason for favouring polluter subsidies; see section 4.4 below.

24 See Ogus and Richardson (1977, section II) and chapter 5 below.

25 See Baumol and Oates (1975, chapter 12) for a resumé of the debate; Kneese and Mäler (1973); Mäler (1974); Kamien, Schwartz and Dolbear (1966); and Bramhall and Mills (1966).

26 In practice there are likely to be considerable difficulties for the agency in attempting to identify this pollution level. If polluters know that their subsidy is to be based on the difference between their pollution levels before receiving the subsidy and after, they will have an incentive to pollute to higher levels than q_n^A to begin with.

27 If each polluter faced horizontal pollution cost curves, on the other hand, the fixed rate charge would lead to socially efficient industry output (pollution) level. See Burrows (1978b); Baumol and Oates (1975, chapter 12).

28 Information on abatement costs is scattered across many sources. Some examples are: Environmental Protection Agency (1972); Kneese and Shultze (1975, chapters 2 and 6); Pearce (1976b, chapter 6 and references on pp. 196–8); White (1976); Burrows, Rowley and Owen (1974b, c); Atkins and Lowe (1977). Two new books, Baumol and Oates (1979) and Pearce (1978a), should provide surveys of data on both pollution and abatement costs.

29 On problems of pollution cost estimation see Pearce (1976b, chapter 6); Kneese and Shultze (1975, chapter 2); Kneese (1977, chapters 2, 3, 7, 8); Lave and Seskin (1974); Ridker (1967); Berry (1977).

30 Pearce (1978a, b).

31 See Baumol and Oates (1975, chapter 6, section 5). On the other forms of collusion which are possible with a charge scheme see Rose-Ackerman (1973, p. 524).

32 See Kneese and Shultze (1975, chapter 7); Baumol and Oates (1975, chapter 10, section 4); Dick (1974, chapter 4); Rowley (1974), to name but a few.

33 E_2^{A+C} was the socially efficient pollution level in the previous analysis. Now we can regard E_2^{A+C} as the overall standard, which may or may not be socially efficient when the MPC curve in figure 4.3 (c) is unknown.

34 An excellent example of this can be found in Baumol and Oates (1975). On p. 140 they say that the abatement cost savings with a charge 'may by no means be negligible', and on p. 15 that abatement can 'frequently be achieved at surprisingly modest cost'.

35 See Kneese and Shultze (1975, chapter 2) and Kneese (1977, chapter 7).

36 One study that did calculate the likely price effects of a regulatory

system for controlling oil pollution at sea owing to deliberate dumping (a major source of such pollution) suggested that the total abatement costs would have a very small impact indeed. Even if it were administratively feasible in this case, a charge would have no advantage. See Burrows, Rowley and Owen (1974b).

37 A synthetic fact is an opinion that one must assert as frequently and as vehemently as possible because
   (a) one's predilections suggest it *ought* to be true;
   (b) one can't actually *prove* it to be true (except perhaps by generalizing from some carefully chosen case studies); and
   (c) constant repetition might insinuate the idea into people's minds so that they will eventually cease to attempt to distinguish it from real facts.

38 See Rose-Ackerman (1973, section III); Tietenberg (1974, 1978); Baumol and Oates (1975, chapter 10, section 5).

39 See especially Tietenberg (1978); Rose-Ackerman (1973, section III).

40 In the short run: in the long run differences arise relating to the different rates of entry of firms under the two policies. See section 4.2.5.

41 See Rose-Ackerman (1973, pp. 522–3).

42 See, for example, Kneese and Shultze (1975, pp. 91–2); Kneese and Bower (1968, p. 136); Baumol (1972b, p. 319). No hard evidence on costs is quoted by these authors.

43 For example, Kneese and Shultze (1975, pp. 89–90).

44 See Burrows (1977, section 2.3).

45 In addition a squeeze on profits may restrict the finance available for research and development which is the major source of current innovation.

46 See Kennedy and Thirlwall (1972) and Shaw and Sutton (1976, chapter 7, section 2).

47 A 'day' in this context is merely some relatively short time period.

48 Baumol and Oates (1975, chapter 11), present a formal model of pollution control with the level of pollution subject to stochastic influences. It seems close to the central point of the analysis, however, to assume instead that it is the damage costs at any level of emission that fluctuate.

49 In saying that *either* a charge *or* regulation is socially efficient in this situation we are implicitly assuming similar individual damage and abatement cost curves for each polluter; this enables us to ignore, for the moment, the other distinguishing characteristics of the two instruments.

50 Baumol and Oates (1975, chapter 11).

51 This strategy could also be applied to pollutants of varying degrees of severity, by setting uniform tight standards for highly dangerous pollutants and allowing the flexibility of response with a charge for less toxic emissions.

52 See Burrows, Rowley and Owen (1974a, section 2) on accidental oil pollution by tankers.

53 An interesting example of the difficulties of bringing home to a tanker company the consequences of its negligence was the Torrey Canyon grounding in 1968. For details see Burrows, Rowley and Owen (1974c).

54 On some difficulties with private law remedies see chapter 5.

55 See Burrows, Rowley and Owen (1974a, section 4) on the post-Torrey Canyon developments in insurance/compensation schemes for oil tanker operators, together with the British legislation in 1971 to require that all oil tankers using British ports carry a certificate of insurance against pollution damage.

56 See Burrows, Rowley and Owen (1974c; 1974a, section 4).

57 For an entertaining, though not very informative, debate on the merits of economic growth which includes reference to environmental concern as elitism, see Beckerman (1971) and Mishan (1972). See also Mishan (1968) and Beckerman (1974).

58 For a more extensive discussion of the theory see Freeman (1972, sections III and IV).

59 This has not prevented many observers from making assertions about the distributive effects of pollution control based on what Pearce calls 'armchair observation'; Pearce (1978b, p.1).

60 Cicchetti, Seneca and Davidson (1969).

61 Freeman (1972, section IV); Berry (1977). However the generalization may be too glib, for Harrison (1975) found that the benefits of the reduction of automobile emissions would not be neatly related (inversely) to income. The explanation given was that low-income groups are dominant in both urban (high benefits from abatement) *and* rural areas (low benefits). Within urban areas, however, the benefits were found to favour the poor.

62 See the survey of the evidence in Pearce (1978b).

63 Gianessi, Peskin and Wolff (1977); Dorfman (1977); and Dorfman and Snow (1975).

CHAPTER 5

1 See Ogus and Richardson (1977, p. 295) on English private law, and Gunningham (1974, chapters 2–5) on legislation.

2 Ogus and Richardson (1977, p. 286).

3 Ogus and Richardson (1977, section 2) provide a comprehensive list of the conditions that must be satisfied.

4 See Ogus and Richardson (1977, section 2B).

5 Ogus and Richardson (1977, p. 309).

6 Morris *v.* Redland Bricks (1970), A. C. 652, 664. However, the

courts will sometimes *delay* the imposition of an injunction to allow the polluter time to comply, perhaps thereby keeping polluter abatement costs down,

7 Temporal separation would be socially efficient if the costs of restricting the activities to certain operating times were low.

8 See Coase (1960, pp. 431–2).

9 Described by Krier (1971, pp. 458–9). The case was Diamond *v.* General Motors Corporation (1969).

10 The following discussion relies heavily on Ogus and Richardson (1978) for the legislative developments in Britain, and on Krier (1974) and Kneese and Shultze (1975) for those in the United States. Further information on British pollution control problems and policy can be found in the six reports of the Royal Commission on Environmental Pollution.

11 See, for example, the Nuisances Removal Act, and the series of Public Health Acts from 1848 to 1936.

12 For references to the reports see Ogus and Richardson (1978, section 3).

13 The United States is considered below. See McLoughlin (1976) on Europe.

14 For example Alkali Acts and the 1956 Clean Air Act.

15 Royal Commission on Environmental Pollution (1976).

16 Frankel (1974).

17 Royal Commission on Environmental Pollution (1976).

18 Richardson, Ogus and Burrows (1980) examine this discretion in some detail.

19 See Gunningham (1974, chapter 6).

20 Krier (1974).

21 See Richardson, Ogus and Burrows (1980).

22 See Kneese and Shultze (1975, chapters 3–6), for a critical appraisal which provides them with a basis for advocating the adoption of charges in place of regulated standards (see chapter 7).

23 An interesting survey of the US legislation and evidence on costs relating to automobile emissions can be found in White (1976).

24 Royal Commission on Environmental Pollution (1974, paras. 48–52 and chapter V).

25 Kneese and Shultze (1975, chapters 6 and 7).

26 Kneese and Shultze (1975, chapter 5).

27 Kneese and Schultze (1975, chapter 6).

28 Kneese and Schultze (1975, pp. 91–2).

References

ATKINS, M. H. and LOWE, J. F. (1977) *Pollution Control Costs in Industry – an Economic Study* London, Pergamon.

AYRES, R. U. and KNEESE, A. V. (1969) 'Production, Consumption and Externality' *American Economic Review* vol. 59, June.

BARKELY, P. W. and SECKLER, D. W. (1972) *Economic Growth and Environmental Decay: the Solution Becomes the Problem* New York, Harcourt Brace Jovanovich.

BAUMOL, W. J. (1972a) *Economic Theory and Operations Analysis*, 3rd edn, Englewood Cliffs, NJ, Prentice-Hall.

BAUMOL, W. J. (1972b) 'On Taxation and the Control of Externalities' *American Economic Review* vol. 62, June.

BAUMOL, W. J. and OATES, W. E. (1975) *The Theory of Environmental Policy: Externalities, Public Outlays and the Quality of Life*, Englewood Cliffs, NJ, Prentice-Hall.

BAUMOL, W. J. and OATES, W. E. (1979) *Economics, Environmental Policy and the Quality of Life* Englewood Cliffs, NJ, Prentice-Hall.

BECKERMAN, W. (1971) 'Why We Need Economic Growth' *Lloyds Bank Review* October.

BECKERMAN, W. (1974) *In Defence of Economic Growth* London, Jonathan Cape.

BERRY, B. J. (1977) *The Social Burdens of Environmental Pollution* Cambridge, Mass., Ballinger.

BRAMHALL, D. E. and MILLS, E. S. (1966) 'A Note on the Asymmetry Between Fees and Payments' *Water Resources Research* vol. II, no. 3.

BROWNING, E. K. (1977) 'External Diseconomies, Compensation and the Measure of Damage' *Southern Economic Journal* vol. 43, January.

BUCHANAN, J. M. (1969) 'External Diseconomies, Corrective Taxes and Market Structure' *American Economic Review* vol. 59, March.

BURROWS, P. (1970) 'On External Costs and the Visible Arm of the Law' *Oxford Economic Papers* vol. 22, March.

BURROWS, P. (1974) 'Pricing Versus Regulation for Environmental Protection' in A. J. CULYER (ed.) *Economic Policies and Social Goals* London, Martin Robertson.

BURROWS, P. (1977) 'Pollution Control with Variable Production Pro-

183

cesses' *Journal of Public Economics* vol. 8.

BURROWS, P. (1978a) 'Controlling the Monopolistic Polluter: Nihilism or Eclecticism?' University of York Economics Discussion Paper 12.

BURROWS, P. (1978b) 'Pigovian Taxes, Polluter Subsidies, Regulation and the Size of a Polluting Industry' University of York Economics Discussion Paper 16, forthcoming in the *Canadian Journal of Economics.*

BURROWS, P., ROWLEY, C. K. and OWEN, D. (1974a) 'The Economics of Accidental Oil Pollution by Tankers in Coastal Waters' *Journal of Public Economics* vol. 3.

BURROWS, P., ROWLEY, C. K. and OWEN, D. (1974b) 'Operational Dumping and the Pollution of the Sea by Oil: An Evaluation of Preventive Measures' *Journal of Environmental Economics and Management* vol. 1.

BURROWS, P., ROWLEY, C. K. and OWEN, D. (1974c) 'Torrey Canyon: A Case Study in Accidental Pollution' *Scottish Journal of Political Economy* vol. XXI, November.

CALABRESI, G. and MELAMED, A. D. (1972) 'Property Rules, Liability Rules and Inalienability: One View of the Cathedral' *Harvard Law Review* vol. 85, April.

CICCHETTI, C. J., SENECA, J. J. and DAVIDSON, P. (1969) *The Demand and Supply of Outdoor Recreation* Washington, US Department of the Interior.

COASE, R. H. (1960) 'The Problem of Social Cost' *Journal of Law and Economics* vol. 3, October.

COLE, H. S. D. *et al.* (1973) *Thinking About the Future: A Critique of the Limits to Growth* London, Chatto and Windus.

DALES, J. H. (1968) *Pollution, Property and Prices* University of Toronto Press.

DAVIS, D. A. and WHINSTON, A. (1962) 'Externalities, Welfare and the Theory of Games' *Journal of Political Economy* vol. 70, June.

DEMSETZ, H. (1971) 'Theoretical Efficiency in Pollution Control: Comments on Comments' *Western Economic Journal* vol. 9, December.

DICK, D. T. (1974) *Pollution, Congestion and Nuisance* Farnborough, Hants, Lexington Books.

DICK, D. T. (1976) 'The Voluntary Approach to Externality Problems: A Survey of the Critics' *Journal of Environmental Economics and Management* vol. 2.

DOLBEAR, F. T. (1967) 'On the Theory of Optimum Externality' *American Economic Review* vol. 57, March.

DORFMAN, R. (1977) 'Incidence of the Benefits and Costs of Environmental Programs' *American Economic Review, Papers and Proceedings* vol. 67, February.

DORMAN, N. S. and SNOW, A. (1975) 'Who Will Pay for Pollution Control?' *National Tax Journal* vol. 28.

EDEL, M. (1973) *Economics and the Environment* Englewood Cliffs, NJ, Prentice-Hall.

ELLICKSON, R. C. (1972) 'Alternatives to Zoning: Covenants Nuisance Rules and Fines as Land Use Control' *University of Chicago Law Review* vol. 40.

ENVIRONMENTAL PROTECTION AGENCY (1972) *The Economic Impact of Pollution Control: A Summary of Recent Studies* Washington, DC, US Government Printing Office.

FLETCHER, G. P. (1972) 'Fairness and Utility in Tort Theory' *Harvard Law Review* vol. 85.

FRANKEL, M. (1974) *The Alkali Inspectorate* London, Social Audit.

FREEMAN, A. M. (1972) 'The Distribution of Environmental Quality' in A. V. KNEESE and B. BOWER (eds) *Environmental Quality Analysis: Theory and Method in the Social Sciences* Baltimore, Johns Hopkins Press.

GIANESSI, L., PESKIN, H. and WOLFE, E. (1977) 'The Distributional Effects of the Uniform Air Pollution Policy in the United States' *Resources for the Future* Discussion Paper D-5, Washington.

GIFFORD, A. (1977) 'Externalities and the Coase Theorem, A Graphical Analysis' *Quarterly Review of Economics and Business* vol. 14.

GIFFORD, A. and STONE, C. C. (1973) 'Externalities, Liability and the Coase Theorem, A Mathematical Analysis' *Western Economic Journal* vol. 11, September.

GOULD, J. R. (1977) 'Total Conditions in the Analysis of External Effects' *Economic Journal* vol. 87, September.

GUNNINGHAM, N. (1974) *Pollution, Social Interest and the Law* London, Martin Robertson.

HARRISON, D. (1975) *Who Pays for Clean Air?* Cambridge, Mass. Ballinger.

HEAD, J. G. (1962) 'Public Goods and Public Policy' *Public Finance* vol. 17, no. 3.

HOLTERMAN, S. (1976) 'Alternative Tax Systems to Correct for Externalities and the Efficiency of Paying Compensation' *Economica* vol. 43, February.

KAMIEN, M. I., SCHWARTZ, N. L. and DOLBEAR, F. T. (1966) 'Asymmetry Between Bribes and Charges' *Water Resources Research* vol. II, no. 1.

KAPP, K. W. (1970) 'Environmental Disruption and Social Costs: A Challenge to Economics' *Kyklos*, reprinted in H. WOLOZIN (ed.) *The Economics of Pollution* Morristown, NJ, General Learning press, 1974.

KENNEDY, C. and THIRLWALL, A. P. (1972) 'Technical Progress, A Survey' *Economic Journal* vol. 82, March.

KNEESE, A. V. (1977) *Economics and the Environment* Harmondsworth, Penguin.

KNEESE, A. V. and MALER, K. G. (1973) 'Bribes and Charges in Pollution ics, Technology, Institutions* Baltimore, Johns Hopkins Press.

KNEESE, A. V. and MÄLER, K. G. (1973) 'Bribes and Charges in Pollution Control: An Aspect of the Coase Controversy' *Natural Resources Journal* vol. 13, October.

KNEESE, A. V. and SCHULTZE, C. L. (1975) *Pollution, Prices and Public Policy* Washington, DC, Brookings Institution.

KRIER, J. E. (1971) 'The Pollution Problem and Legal Institutions: A Conceptual Overview' *UCLA Law Review* vol. 18.

KRIER, J. E. (1974) 'The Irrational National Air Quality Standards: Macro- and Micro-Mistakes' *UCLA Law Review* vol. 22.

LANCASTER, A. (1966a) 'Change and Innovation in the Technology of Consumption' *American Economic Review Supplement* May.

LANCASTER, A. (1966b) 'A New Approach to Consumer Theory' *Journal of Political Economy* vol. 74, April.

LANCASTER, A. (1969) *Introduction to Modern Microeconomics* Chicago, Rand McNally.

LAVE, L. B. and SESKIN, E. P. (1970) 'Air Pollution and Human Health' *Science* vol. 169, August.

LAVE, L. B. and SESKIN, E. P. (1974) 'Does Air Pollution Shorten Lives?' in J. W. PRATT (ed.) *Statistical and Mathematical Aspects of Pollution Problems* New York, Marcel Dekker, 1974.

LEE, N. and SAUNDERS, P. J. W. (1972) 'Pollution as a Function of Affluence and Population Increase' in P. R. COX and J. PEEL (eds) *Population and Pollution* New York, Academic Press, 1972.

LIPSEY, R. G. (1975) *An Introduction to Positive Economics* London, Weidenfeld and Nicolson.

MÄLER, K. G. (1974) *Environmental Economics* Baltimore, John Hopkins Press for Resources for the Future.

MARCHAND, J. M. and RUSSELL, K. P. (1973) 'Externalities, Liability, Separability and Resource Allocation' *American Economic Review* vol. 63, September.

McLOUGHLIN, J. (1976) *The Law Relating to Pollution: An Introduction* Manchester University Press.

MEADOWS, D. *et al.* (1972) *The Limits to Growth* London, Earth Island.

MILLWARD, R. (1971) *Public Expenditure Economics* London, McGraw-Hill.

MISHAN, E. J. (1967) 'Parento Optimality and the Law', *Oxford Economic Papers* vol. 19, November.

MISHAN, E. J. (1968) *The Costs of Economic Growth* Harmondsworth, Penguin.

MISHAN, E. J. (1971) 'The Post-war Literature on Externalities' *Journal of Economic Literature* vol. 9, March, reprinted in H. WOLOZIN (ed.) *The Economics of Pollution* Morristown NJ, General Learning Press, 1974.

MISHAN, E. J. (1972) 'Economic Growth: the Need for Scepticism' *Lloyds Bank Review* October.

OGUS, A. I. and RICHARDSON, G. M. (1977) 'Economics and the Environment: A Study of Private Nuisance' *Cambridge Law Journal* vol. 36, November.

OGUS, A. I. and RICHARDSON, G. M. (1978) 'The Regulatory Approach to Environmental Control' paper presented to CES Conference on Urban Law at Queen's College, Oxford, June 1978 (mimeo).

PEARCE, D. W. (1976a) 'Environmental Protection, Recycling and the International Materials Economy', in I. WALTER (ed.) *International Economic Dimensions of Environmental Management* New York, John Wiley, 1976.

PEARCE, D. W. (1976b) *Environmental Economics* London, Longman.

PEARCE, D. W. (1978a) *The Valuation of Social Cost* London, Allen and Unwin.

PEARCE, D. W. (1978b) 'The Social Incidence of Environmental Costs and Benefits' *University of Aberdeen Occasional Paper (Economics)* No. 78–08.

PORTES, R. D. (1970) 'The Search for Efficiency in the Presence of Externalities' in P. STREETEN (ed.) *Unfashionable Economics: Essays in Honour of Lord Balogh* London, Weidenfeld and Nicholson, 1970.

POSNER, R. A. (1972) *Economic Analysis of Law*, Boston, Little, Brown.

RAWLS, J. (1971) *A Theory of Justice* Oxford University Press.

RICHARDSON, G. M., OGUS, A. I. and BURROWS, P. (1980) *Administrative Discretion and the Enforcement of River Pollution Law* London, Macmillan, in press.

RIDKER, R. B. (1967) *Economic Costs of Air Pollution* New York, Praeger.

ROSE-ACKERMAN, S. (1973) 'Effluent Charges: A Critique' *Canadian Journal of Economics* vol. 6, November.

ROWLEY, C. K. (1974) 'Pollution and Public Policy' in A. J. CULYER (ed.) *Economic Policies and Social Goals* London, Martin Robertson, 1974.

ROYAL COMMISSION ON ENVIRONMENTAL POLLUTION (1972) *Third Report: Pollution in Some British Estuaries and Coastal Waters* Cmnd 5054, London HMSO.

ROYAL COMMISSION ON ENVIRONMENTAL POLLUTION (1974) *Fourth Report: Pollution Control: Progress and Problems* Cmnd 5780, London, HMSO.

ROYAL COMMISSION ON ENVIRONMENTAL POLLUTION (1976) *Fifth Report: Air Pollution Control: an Integrated Approach*, Cmnd 6371, London, HMSO.

SAMUELSON, P. A. (1954) 'The Pure Theory of Public Expenditures' *Review of Economics and Statistics* vol. 36, November.

SENECA, J. J. and TAUSSIG, M. K. (1974) *Environmental Economics* Englewood Cliffs, NJ, Prentice-Hall.

SHAW, R. W. and SUTTON, C. J. (1976) *Industry and Competition* London, Macmillan.

SLATER, M. (1975) 'The Quality of Life and the Shape of the Marginal Loss Curves' *Economic Journal* vol. 85, December.

STARRETT, D. and ZECKHAUSER, R. (1974) 'Treating External Diseconomies – Markets or Taxes?' in J. W. PRATT (ed.) *Statistical and Mathematical Aspects of Pollution Problems* New York, Marcel Dekker, 1974.

TIETENBERG, T. H. (1974) 'Taxation and the Control of Externalities: Comment' *American Economic Review* vol. 64, June.

TIETENBERG, T. H. (1978) 'The Quasi-Optimal Price of Undepletable Externalities: Comment' *The Bell Journal of Economics* vol. 9, Spring.

VICTOR, P. A. (1972) *Economics of Pollution* London, Macmillan.

WELLISZ, S. (1964) 'On External Diseconomies and the Government – Assisted Invisible Hand' *Economica* vol. 31, August.

WHITE, L. J. (1976) 'American Automotive Emissions Control Policy: A Review of the Reviews' *Journal of Environmental Economics and Management* vol. 2, April.

Author Index

189

Subject Index